T0205854

Reflections on Slope Stability Engineering

Slope stability engineering is both an art and a science practised by many civil and geotechnical engineers involved in work on landslides, earth dams, breakwaters, coastal defence, earthworks slopes for roads, canals, railways, pipelines, housing developments, and other related developments.

This book contains the detailed reflections of its author, who has practised and researched in the field for over a half-century. It is written in an informal style that makes it an interesting and thought-provoking practitioner guide to landslides and slope problems and their investigation, analysis, and remediation, considering both natural and man-made slopes and earthworks, and without the need for the usual equations and illustrations.

Reflections on Slope Stability Engineering is targeted primarily at practitioners working in the investigations of slope instability and the design and construction of treatments of the problem, especially those early in their careers, but the accessible style also suits students who are developing an interest in the subject and even those engineers with only a casual interest in this branch of geotechnics.

Edward N. Bromhead is a retired Professor from Kingston University, UK, and a former consulting engineer. He is a Chartered Civil Engineer, a Fellow of the Geological Society, a former Glossop Lecturer and was awarded The Varnes Medal in 2023. He is author of *The Stability of Slopes* (CRC Press) and many journal and conference papers.

Reflections on Slope Stability Engineering

Edward N. Bromhead

CRC Press is an imprint of the
Taylor & Francis Group, an **informa** business

Cover image: Edward N. Bromhead

First edition published 2024
by CRC Press
4 Park Square, Milton Park, Abingdon, Oxon, OX14 4RN

and by CRC Press
6000 Broken Sound Parkway NW, Suite 300, Boca Raton, FL 33487–2742

CRC Press is an imprint of Informa UK Limited

British Library Cataloguing-in-Publication Data
A catalogue record for this book is available from the British Library

ISBN: 978-1-032-54931-6 (hbk)
ISBN: 978-1-032-54895-1 (pbk)
ISBN: 978-1-003-42816-9 (ebk)

DOI: 10.1201/9781003428169

Typeset in Sabon
by Apex CoVantage, LLC

Contents

Preface

THIS BOOK

This book owes its origins, in part, to my two-day course on slope stability that I ran for many years, initially for ICE training, or as it used to be called, Thomas Telford Training, and latterly for Synergie Training. I was reluctant at first to run the course, as it was at the time when such courses demanded a huge repertoire of PowerPoint presentations, and although I had amassed a great many 35mm slides and transparencies for the overhead projector, it looked like a big job to prepare. In the event, I found that it was less of a chore than I feared, as I had to move just about everything I had to PowerPoint anyway for teaching and research reasons long before I retired from Kingston University.

I had no intention of duplicating the efforts that went into my book on the *Stability of Slopes* and therefore wrestled for a while to find an appropriate format. The answer, I think, that has given rise to this book, again, has multiple origins. Both my home and my home office have shelves that are groaning with books and files. I've always been an avid reader, and so it isn't just technical stuff but also a collection of recreational reading items dating back to my boyhood. When I retired from Kingston University, the act of clearing my office added to the stress on my 'library space' at home. At about that time, I acquired a Kindle e-book reader. The e-book format lends itself to a book without illustrations or formulae, and the question for me was "*Could I write a book on a technical subject without using figures or equations?*" I had a go, and this book is the result. It occurred to me that such a book answered the question: "*How do I get to the really important matter without being diverted by trivia, and build on the basic materials that all potential readers will have covered at least once in college or University*".

Part of the decision to forego figures and equations lies in the fact that most readers of a book like this will have had access to those sorts of diagrams and equations in their undergraduate or postgraduate studies. What the book needs to do is to simply cause the reader to conjure up the relative figure from their education or their professional work to match what

I am telling them. It's something like the experience that I get when I read a novel on my Kindle. I am a particular fan of the novels written by the Italian author Andea Camilleri about the cases of Inspector Montalbano in the fictional town of Vigata in Sicily. Whenever Montalbano dines out, I see in my mind's eye the sort of restaurant that I have visited many times in Italy. I can conjure up a view of the weather patterns from the briefest descriptions. I don't always see the characters or sets from the television series, but I get a real feel for Italy that is quite good enough. The books aren't even in the original Italian: I read the translations by Stefan Sartorelli. And every time the author mentions dodgy buildings that fall down, towns that slide downhill, and the politics of Italian life in the south of the country, I can relate to my own experiences there. I don't need the book to be illustrated, nor do I need to be told the minutiae of Italian Law to understand the predicaments that Montalbano sometimes finds himself in, or the rivalry between the Municipal Police and the Carabinieri, since there is similar competition both in Engineering and in Academia.

To test the validity of the assumption that a technically qualified reader can mentally conjure up a picture, now think about a simple earth slope, with a slip circle on it. Divide the sliding mass into vertical slices. Now add some information about the water conditions in the slope, and how the slope is divided into different materials. I'll bet that you could draw it, too. Now, did you really need me to do it for you? And did you really need me to list, or develop, the equations in the software that you will use in the computer application you will use to analyse it? Aren't they in the user manual, or papers that are now accessible online? Or even in my other book?

So, this is a book about my thoughts and work in slope stability engineering for you, the reader, to set within the framework of the education and experience you already possess. It is my sincere hope that you will sometimes find nuggets of wisdom, but equally, I want you to nod and think, "*Ah, yes. That's why we do such-and-such. It makes sense now*" or "*Yes, I agree. We aren't doing that well enough. I'll give it some attention*". I have tried to write it in the grammar of speech, rather than the strict grammar of academic writing, and hope that my colloquial mannerisms, jibes, and occasional attempts at humour do not grate on your nerves.

The other thing that I have taken from the translated Montalbano books is the use of endnotes. Not, as in those books, endnotes for the whole book, but chapter by chapter. These end notes are not just simply academic journal references, they sometimes contain information that might break the flow of the text or make it overlong. I have hopes that this book will be read in a traditional form, *i.e.* in a codex (a *book* being either that or a scroll), as well as on an e-book reader. I am certain that I will have proofread it on my Kindle, as I find that transferring to that much smaller page makes it easier to find mistakes, particularly typographic errors, although I have to admit that the copy editor found many. In part, the endnotes are answers that I might give to some questions by someone interrupting my flow.

I have another wish, too, and that it is that the professional world finds my rationale for adopting this format a persuasive one, and that this book becomes the first in a series, all by different specialist Authors, with titles that begin: '*Reflections on . . .*' perhaps with the implicit sub-title: '*Core Principles for Civil and Geotechnical Engineers*'.

SLOPE STABILITY AND LANDSLIDES

All the way through this book, I have written about landslides, how to stop them from occurring in the first place, or to stop them in their tracks if you are already faced with a landslide or landslides on your site. That begs the question of what landslides are, or even why the word 'landslide' is a sensible one to use – or even if slope stability engineering is all about 'landslides'. Slopes and earthworks settle, or sometimes heave quite independently of whether they will suffer the sort of movements that the word 'landslide' describes, but those other types of movement can be treated separately, and so for the purpose of this book, slope stability engineering is all about 'landslides'. It turns out that 'landslide' is not even the most favoured word for some interested parties, and indeed, 'landslip' was the more common word in English until some way into the 20th Century – it is still the favoured word in railway engineering.

Throughout the book I have used, by habit, the phrase *Factor of Safety* using capitals to show how important it is. The phrase denotes more or less how far away from collapse a slope is, or what reserve of strength there is to stop a landslide from moving. Sometimes, I just abbreviate it to the symbol *F*. A pet hate of mine is the halfway house, *FoS*.

ME, THE AUTHOR

Some readers will know me. I normally use the diminutive of my first name. I am a graduate of the forerunner of the University of Portsmouth, where I studied a BSc degree in Civil Engineering in the 1960s. I took the MSc course in Soil Mechanics at Imperial College in the early 1970s, and at the end of the decade and into the next studied part-time there for my PhD, the latter done when I was a lecturer at what is now Kingston University. I worked in industry at first, originally with ambitions to be a structural engineer, but I had a difference of opinion with the boss, and my next job was in a geotechnical section, where dealing with slope stability caught my imagination. I became a Chartered Civil Engineer in 1974, having been both a student member and a graduate, so my 50 years certificate is now a little old, and I have continued to do research and publish papers since retirement from academic life – and do work on books like this one.

I found that the academic life suited me because of the freedoms it gave. I found time to do consulting and develop my research interests in the field, programming on mainframe computers and later a PC, and doing testing in the geotechnics laboratory including on apparatus I designed myself. The book is informed by over a half-century of work, some at the cutting edge of technology, some rather more mundane, but all of it, in my experience, thought-provoking. I often think of myself as working alone, that is in the sense of working in a room with the door shut, but as far as a project goes, I am always part of a team. Since I moved into the world of geotechnical engineering, I have learned from many different individuals, some of whom have been in a formal teaching relationship, others just people I have met semi-socially and who have shared interests together with others met while 'on the job'. In those jobs, incidentally, I have worked on coastal landslides, inland landslides, mountain landslides, dams and dam sites, static and seismic problems, drawdown and submerged slopes, embankments and cut slopes for roads, railways, canals, and pipelines. In those jobs I have done investigations, testing, analyses, design of remedial works, often using software I wrote myself or apparatus (such as my small ring shear device that was put into production by Wykeham Farrance Engineering) that I designed, built and developed. I have other skills, too. I have given evidence based on my work to Mediations, Arbitrations, and in the High Court, and also to a Parliamentary Committee considering a Private Member's Bill. Add to that the amount of instrumentation and field testing I have done, and I think that it isn't just *chutzpah* to think that I am a suitable author for a book like this.

However, I have limitations. After I left industry, I stopped doing a lot of work on foundations, and although I have drilled soils and rocks, using augers, or rotary coring (with air, water, and synthetic polymer flush) I've never been deeper than 85m with me at the controls of a drilling machine!

Many of those who taught or helped me are, sadly, no longer with us, but I still learn from my contemporaries, and am prepared to learn from anyone I encounter. I have particularly fond memories of the academic staff at Portsmouth, Imperial College, and my colleagues and students at Kingston, together with those individuals, too many to mention individually, who I have worked with in research and practice over the years.

If I can say what the greatest revelation in my professional life has been, it is the finding of how important geology is. It should therefore come as no surprise that as well as being a Chartered Civil Engineer, I am a Fellow of the Geological Society of London, that my first scientific paper appeared in the *Quarterly Journal of Engineering Geology*, or that I spent time as Editor of that Journal after it had added *'and Hydrogeology'* to its name. It is the responsibility of senior members of any profession to share their experiences and help colleagues develop themselves professionally. I hope that this book fulfils some of that duty.

I have some simple pleas to you, the reader, and they are that when you are in the field, undertaking investigations or doing remediation works, you pay great attention to your own safety, and the safety of those around you. I have been in the office when news of a colleague's death on site was announced, and I have even turned up on to a site where an engineer was killed in a landslide shortly before I arrived. The safety of those around you is as much your responsibility as your own safety. This isn't a book about safety on site, but aiming for a safe and healthy working environment should be an aim in everything you do.

Anyone can ask me about anything in my experience. But please, don't ask me, even if you are a fellow Bromhead, "*Are you (or we) related to the bloke played by Michael Caine in the film* Zulu*?*" Not unless you've just bought me a drink and it's a social occasion, and in addition, you are prepared for a long and sometimes boring exposition on the subject of the Zulu War.

Edward (Eddie) Bromhead
Yateley, March 2023

Chapter 1

Landslide recognition, description, classification, and mapping

WHAT IS A LANDSLIDE?

Not all slope stability engineering is concerned with landslides: some of it is about preventing landslides, not just controlling or stabilising them, and some instability mechanisms don't involve landslides, but most do.

Well, what *is* a landslide? Why is it important to understand what causes landslides? How do we investigate landslides? How do we fix the underlying problems? This book attempts to answer these and related questions in as simple a way as possible for a readership of geotechnical professionals, particularly geotechnical engineers and engineering geologists. It's a very personal account of over a half-century in the field as a practitioner, a researcher, and a teacher in Higher Education, and, since I retired from Kingston University, through the mechanisms of the training short courses that I have run, various consulting jobs, speaking engagements, and in reading and reviewing papers written by others.

To return to the first question, what is a landslide? We use the word 'landslide' in common speech today mostly to denote some political upheaval in terms of an overwhelming majority of votes for one party or candidate in an election that is, up to a point, unexpected but always disastrous for the loser. Normally, it is '*won by a landslide*', but not '*lost by a landslide*'.[1] That isn't the sort of landslide that this book is about. No, here we are dealing with something else. You can go to various places for a definition, including Wikipedia, which describes it as a form of *mass wasting*, which shows that the page wasn't written by an engineer, as one is likely to find in a dictionary of geology or geography that mass wasting *includes* landslides, the sort of circular definition that I once found when trying to discover what a *gazebo* was, as at the time, I wasn't sure. This particular gazebo in Tenby, Pembrokeshire, had collapsed down a cliff under the enthusiastic shaking of a courting couple, killing them both. My dictionary told me that a gazebo was a *belvedere*, and being no wiser, I looked and found that a *belvedere* could be a *gazebo*! But it was involved in a landslide, and it caused fatalities, and it was a form of human-induced landslide, albeit an unusual seismically induced one.

DOI: 10.1201/9781003428169-1

My preferred definition is that a landslide is a mass of soil or rock that detaches from the parent rock or soil mass, and subsequently moves largely independently before coming to rest. The *moving independently* bit excludes other forms of ground movement such as ground subsidence due to fluid abstraction, volcanic inflation, fault displacement, and a myriad of other ground movements that are largely movements of the parent rock or soil mass, perhaps involving rupture but not to the extent of creating an independently moving mass. I would also include the same phenomena affecting engineered or built slopes of soil, so that isn't mass wasting either. The main geomechanics societies support a Joint Technical Committee (JTC-1), whose remit is *'Landslides and engineered slopes'*, the successor to TC11 of the ISSMGE, which adopted the same remit, which is probably where I get the scope from, as I was a member of both committees in turn. Perhaps the Committee's title should have included '. . . *badly* engineered slopes'.

Of course, other words are used: 'landslip', for one. The word often suggests smaller or perhaps more gradual movements than the word 'landslide'. Other terms include 'slope failure',[2] 'slope movement', the dreadful 'mass wasting' and the similar but slightly less ponderous 'mass movement', 'gravity displacement', the horrid 'slump' and the antonyms 'slope stability' and 'slope instability'. 'Slump' for me will always mean that test which every fledgling civil engineering student does in the concrete laboratory using a truncated cone full of wet concrete to measure its workability. I pity those who do such testing for a living. Some slopes indeed may slump in the same way as that concrete does, for example, an embankment[3] built from material with a low shear modulus, but that is bad slope stability engineering, not a landslide.

As well as the political context, we have to remember that the word 'landslide' is employed even in geotechnics in two contexts. First of all, it is an *event*, particularly when the movement is rapid or violent, which means that it is easy to associate a time and date to it, for example, 'the Aberfan landslide of 21st October 1966' or 'the Vaiont landslide of 9th October 1963'. The other usage of the word denotes the post-event *landscape feature*, such as in 'the coastal town of Ventnor, Isle of Wight, is built mainly on an old landslide'. The *landscape feature* usage doesn't rule out the possibility of future events, and sometimes, if the landslide is sufficiently complex, it is recognised as being the result of multiple landslides by the use of the plural.

Landslides can range in size from the annoying, but far from life-threatening, to colossal, unstoppable, devastating or overwhelming things that can destroy whole cities. Such landslides at all scales are inevitably costly to fix, even if they just need tidying up. The cost of collateral damage can be many times more than the simple clean-up cost, so clearly, we don't want landslides in the first place. Sometimes, in seismically active parts of the world, the landslides are seen as adjuncts to earthquakes, with their deaths and destruction attributed to the earthquake rather than to the resulting landslides. Under seismic shaking, even flat ground can be ruptured so that a body of material moves independently, which is why I sometimes prefer to avoid the use of 'slope' in the definition. I suppose, too, that I ought not to like 'landslide' because an undersea landslide isn't 'land', and a landslide which flows isn't sliding, but there are limits to how fussy one should get in subdividing what is largely a spectrum into a set of classes.

The technology of fixing landslide problems is the business of geotechnical professionals, to whom this book is addressed, and before landslides can be fixed, they need to be investigated and understood, and the remedial work quantified by means of appropriate analyses. This procedure is not normally the job entirely of a one-man band, so a common language is required for communication within a team. That means differentiating between certain kinds of landslides, but not an over-elaborate one in which every landslide or landslide-like feature has its own class.[4]

IS IT A BIRD, IS IT A PLANE?

OK, we've decided on what a landslide is, but can we be more specific? I'm all for simplification, and the scheme I like best can be simplified even further. It is a scheme invented by Cruden and Varnes,[5] and in a simplified form, it was used as a framework for a European-funded collaborative project that produced a book[6] with examples of every landslide type, mainly from Europe. The scheme relies on two parts: a *materials descriptor* and a *characteristic mechanism*. The *materials descriptor* is used in a qualifying or adjectival sense to the *characteristic mechanism*, and may be one of three things: *rock, soil,* or *debris*. Some people have difficulty with the idea of a soil composed of boulders, but boulders enter into most soil classification schemes, so I do not. However, since the result of earlier landslides may be an accumulation of boulders large and small, and people do have difficulty with this as soil, again at a human scale, the word 'debris' has to be allowed, and it is also a nod to previous usage – as is also the equivalence of *'earth'* for *'soil'*, differentiated from rock by means of a simple explanation that I will give a page or so on from here.

The characteristic mechanisms include three that geotechnical specialists are most likely to encounter, a couple that only occur in specialist terrain, and one that is, in many cases, a landslide in course of development rather than one that is fully fledged, that is when it is completely detached, and which is moving independently.

The three that we do meet commonly are given the names *slide*, *fall,* and *flow*. The distinguishing characteristic of a *fall* is that the rocks or soils become airborne at some stage. Major falls break up into many independently moving fragments, sometimes as a result of impacts with the parent rock mass, and sometimes with each other. If there are enough fragments, they turn into a *flow* and tend to hug the terrain slope unless they are launched off a cliff. The primary characteristic of a *flow* is that it involves many particles or fragments that mix up as they move, while its secondary characteristic, shared with a *slide*, is that the main landslide body stays largely in contact with the parent rock mass as it moves. *Slides*, on the other hand, comprise one or a few masses in the landslide body, that conceptually might be pushed back into their original location, although that would be a paper exercise, and not a practical proposition in the field.

Typical combinations of materials descriptor and characteristic mechanism might, then, most commonly include *rock fall, soil slide* and *debris flow*, with *rock flow* being perhaps the least likely of the nine combinations.

The two types of characteristic *mechanisms* that are less likely to be encountered in routine practice are the *sag* and the *spread*. Commonly, these are known as 'sagging' or 'spreading', and even then, the German word *Sackung* or its plural *Sackungen* is used for a sag, and spreads are unnecessarily further elaborated with 'lateral', as in 'lateral spreading'. Sags are basically the appearance of normal faults high up in mountains, and really indicate that there is a big slide in progress, one that is inhibited from expressing itself because there is something obstructing its toe like another mountain. These features are on such a scale that no-one will sink a borehole to investigate them, and although they are liable to move so that they even show up through a fresh snowfall, there is usually little chances of them developing completely. On the other hand, using their own nomenclature, Eisbacher and Clague[7] refer to quite ordinary slopewash 'sagging' down the valley side, which, as far as I am concerned, is a good enough description of the process, and perhaps better if one avoids 'Sag' altogether and substitutes 'Sackung'.

Spreads are common in weak alluvial materials such as quick clays[8] that only occur in places like Scandinavia and Eastern Canada. They appear to me to be slides developing into flows. Several mountain features described as 'rock spreading' aren't convincing to me. I also see references to 'deep seated gravitational displacement' (DSGD), which looks to me a lot like tectonics – or movements of the parent rock mass – and therefore not landsliding. Hills or mountains containing a syncline are common, and the syncline usually isn't because the upland sank in the middle.[9]

Then, we come to the bizarre. The term 'flowslide' crept in from somewhere to describe the movements transitional between slide and flow, like the landslide at Aberfan in 1966[10] and other disastrous slides turning into flows. The composite word 'slideflow' that properly reflects the sequence doesn't so easily trip off the tongue. When used, flowslides typically have a long runout. Many landslides would fit into the category of a slide turning into a flow, as you will see if you search the internet.

The oddity is 'topple', which, like 'spread' and 'sag' is normally used ending in 'ing'. Toppling occurs when a weak underbed is eroded or excavated or it is crushed by something more competent and heavier above, allowing that material to rotate out of the slope, *i.e.* by rotation about a centre inside the slope. It is a characteristic behaviour of steep, often vertical, slopes and is much more common in rocks than soils. The soil toppling behaviour is often seen in vertical riverbanks and trench sides, whereas rock toppling occurs in cliffs or, less commonly, in the rear scarps of big landslides. If the weak, erodible or crushable material is at the bottom of a cliff, the toppling block or blocks may rotate considerably, sometimes detaching completely and falling flat. When they do that, they are properly landslides, but while incompletely developed, they are landslides in the process of development.

Rather interestingly, the landslide that perhaps has been seen by more people than any other is a topple. What is it? It's the Pride Rock in the *Lion King* films, which is a slab that has separated along one of a set of major discontinuities and which has toppled off the side of an inselberg.

DESPISED OR DEPRECATED TERMINOLOGY

Wet clayey soils are usually described as mud, and mud has been used as a prefix from time to time. Hutchinson[11] wrote two papers in which he used the term 'mudflow' for a flow-like landslide type in high-water-content London Clay coastal landslides in north Kent. Subsequently, through a field experiment to determine the velocity distribution at the edge of one such landslide, he discovered that it moved as a solid plug. He also discovered that the moving mass slid along a basal shear surface. This was enough to make him embark on a crusade with missionary zeal to rename them all *mudslides*, although he sadly remarked to me once that he encountered obstructions, particularly from North Americans who regard the word 'mudslide' with some horror, as it is a tabloid newspaper catch-all for any landslide disaster – a bit like the effect on me of the word 'slump'. I was shown a webpage of one such US 'mudslide', which contained rocks that, if not exactly the size of houses, certainly dwarfed one of those removal van 'pantechnicons' (as we called them when my father was posted to yet another army camp and my family moved yet again).

Then, you have the highly specific names. One is, of course, 'flowslide'. Another is the 'quick clay landslide'. Both of these are slides turning into flows. And in what way is *liquefaction*, a *lahar* or a *sturzström* not a flow?

I dislike terms such as 'mass movement' and 'mass wasting' applied to landslides as generic terms, when the word 'landslides' will do, even though I am happy for them to be applied to the general geomorphological processes by which landscapes evolve. I detest 'slip plane' because although some slip surfaces are planar, generally, they are not, and some authors even refer to 'circular slip planes'. The term 'slip surface' is not one I particularly like, as it can be either the feature developed at the junction of a landslide and the parent rock or soil mass, or it can be that notional surface along which failure might just occur, considered during a stability analysis procedure (Chapter 4). Alternative terms include 'shear surface', 'basal shear' or just 'shear', 'sliding surface', and when they follow a thin bed or other structural feature that has been termed a 'slide prone horizon' or 'SPH'.[12] I tend to use 'slip surface' when referring to the whole thing because of long-standing habit and not because of any logical preference.

While I am at it, I might add that since this is a book about slope *engineering*, there are landslides of such a scale or in such a place that we really cannot do anything about them, but there is some sort of a competition to describe 'the world's biggest palaeo landslide' or 'the world's biggest submarine landslide' or even 'a huge landslide on Mars'. Books have been written that are compendiums of the biggest and most disastrous.[13] The worst of it all is that it is tempting to join in the game! I am afraid that landslides of this type have (almost) no place here, the exception being those huge landslides that block rivers, which I will discuss later, in the following chapters.

The other difficulty one has when working internationally is translation back and forth, where the origin language has a word not present in the other, so that when something is described by engineer A, engineer B with a different language translates it into his native tongue, and then decides that A's description was wrong. Or engineer B translates his terminology into A's language with a dictionary, and comes up with something not just wrong, but also slightly silly, as in the case of the Chinese author who came up with 'turfy soil' when describing fibrous peat. The fact that at that time, there was no 'turfy soil' on the internet[14] convinced him that he had made a ground-breaking discovery. He still won't be told. Then, finally (for now) the difficulty that speakers of the numerous Latin-based languages have in accepting that words originating in their cultural milieu have been imported into English as loan words, and therefore have subtle differences in meaning from how they use them. And don't get me started on the differences between UK (International!) English and American – that's not one of my 'favorites'!

ROCK, SOIL, AND EARTH

And what is rock, and what is soil? To a geologist, it is all rock, and if it's really weak or recent, then it's 'gardening', not geology. To the civil engineer, however, there is a relatively clear distinction, and that's how you dig it out, rock being more difficult and therefore more expensive, and, as a result, carrying a Bill of Quantities' price premium relative to soil. By this definition, rock means material that needs to be broken up first, by hand using a pick, explosives, wedges, or sometimes drillholes filled with expanding grout, but soil can be dug with a spade, or if especially loose, by a shovel. Not that many people in the developed world would use hand tools given the choice, but much historic infrastructure was dug by hand, and there are still parts of the world where human labour in vast quantities is used, as much to provide work as to save money. The relief of unemployment was a factor well into the 20th century in Britain, and may well be so elsewhere even in the present day.

I have done most of my work, but not all, on soil landslides, and it will usually be the case that I use the words 'soil' and rock' without much differentiation unless it is clear from the context. The big differentiation at the excavation stage is that soil is often weak enough to fail by shear through the intact material, but rock more usually fails along or through pre-existing discontinuities such as joints or faults, or along bedding. However, even in soil, instability seeks out weak zones, that are sometimes depositional in character, and mountain landslides prove quite capable of shearing through the rock mass, so the differentiation is at least in part the result of considering things at the human scale.

WHAT LANDSLIDES CAN WE DEAL WITH?

It's a sad fact that a rapid-onset or fast-moving landslide gives us little option except to get or keep out of the way until it's over. Some forms of rapid-movement landslides can be deflected, or if small enough, be caught, but this requires the foresight or experience to know the routes for rockfalls and debris flows. Topples are a problem mainly where they turn into falls or open cracks at their heads. The main responses to landslides fast or slow are to stabilise the slope before a landslide occurs, or to lock it in place (*i.e.* stabilise it) if it is moving slowly enough. These means are described later on in the book, in chapters on treatment both with and without stabilisation, Chapters 8 and 9.

However, it does mean that for the above reasons, landslide treatment is normally carried out on slopes that are just about stable or just about unstable, where 'unstable' also means an actual or potential landslide we have predicted will move in response to some action, such as digging away at it, it being subject to heavy rain or an earthquake. 'Unstable' is a term that therefore covers *potentially* unstable as well as *actually* unstable and therefore moving.

The fact that working on moving landslides is difficult or dangerous also means that investigating landslides is done on marginally unstable slopes, and landslide recognition is done on the landscape features and not the events, or landslides that have come to rest. As even fast flows can come to rest with a terminal slide, when stationary, they look like old slides, and the only sign that they were flows may be in the shape of pressure ridges in the topography, and the textures in the soil revealed in an excavation.

Mostly, the landslides that slope engineers deal with after movement has taken place are slides, and these have a nomenclature all of their own based on shape. Slides that are long but narrow are sometimes described as *lobate* (tongue-like) with the accumulation zone referred to as a *lobe*. Such slides are similar to flows, following the terrain slope. Deeper-seated landslides are usually *rotational* in whole or part, while a slide that has a rotational part and a structurally controlled planar part is referred to as *compound*. The rotation is about a notional point up in the air above the slope, leading to a curvature in the shape of the slip surface that results from the stress field in the ground. The planar part of a compound slip surface is usually where sliding occurs along a stratigraphic feature such as a thin weak bed, or a thrust fault.

INVESTIGATING LANDSLIDES AND SLOPES

The call comes: to investigate a recent landslide. Or to visit a site, and report back on its landslide issues. Or, more difficult, to identify landslide problems at a site where they are unknown. The techniques are similar between all three; just the amount of experience and knowledge that you have to bring to bear are different. In the following, I describe some of the differences in approach for these three scenarios.

It is worth remembering, of course, that this book is about slope stability *engineering*, and in the investigation phase that means not only examining, describing, classifying, and reporting the occurrence of landslides in an academic sense, but, during a site visit, to be aware of the needs for access for the equipment and personnel to do a ground investigation (Chapter 2). Where, for instance, are we going to position the site offices and welfare facilities? This site accommodation may require little more than site sanitary facilities, but for large projects, it might include canteens, a first aid post, administrative and other offices, together with buildings or shelters in which to store or log samples, and site security facilities to eliminate the worst effects of theft or vandalism. In remote places, the site accommodation may need to include dormitories.

Then, in due course, if stabilisation turns out to be the way forward, it may be necessary to explore access for construction plant, sources for materials or places to dispose of unwanted soil or rock. Investigations of the site and its surroundings inevitably takes on matters such as ownership of neighbouring land, whether or not the site or its flora, fauna or indeed geology are protected in some way or how safe working on site may be affected by tides, river flows, weather or other uncontrollable factors. There are locations in the world where the wild animals or the inhabitants are aggressive and dangerous, and other dangers come from plants and minerals. Be prepared!

THE VISIT TO THE SITE OF A LANDSLIDE DISASTER

It is a really tough brief to be asked to visit the scene of a disaster, especially if there have been fatalities. If there is a rescue in progress, you should keep well clear unless you are an accredited member of the rescue team or a qualified first aider. In any case, feelings will run high, and you must take pains not to be mistaken for a sightseer, or worse these days, a member of the press. Frankly, without the position and training to contribute usefully, you are better off directing traffic or making tea.[15] The only point in a geotechnical professional attending during any rescue is to gather evidence about what happened, and that, up to a point, is getting in the way.

All the rescue workers will be uniformed, and construction worker volunteers will be dressed in high-visibility clothing and helmets, and this combination usually deflects the anger of people affected directly and is therefore the sort of minimum that should be worn. I recommend full personal protective equipment (PPE), as that is in any case part of your toolkit (Chapter 10).

You should remember that many types of disastrous landslides involve secondary failures, and so making an early site visit is potentially very hazardous. If you arrive by car, you should not obstruct emergency vehicles, and park well away from the footprint of the landslide. You should adopt proper safety precautions and abide by company risk assessment rules, as well as instructions from the director of any rescue mission.

It is much more likely that a mission to visit the site of a disaster will take place after the rescue work has been carried out, but there may still be neighbours and relatives in the vicinity who are in a great state of shock and grief, and again, it is necessary to dress recognisably in site clothing, and to conduct yourself with due sensitivity to the situation. If you do not speak the language of the people affected, you must take an interpreter. It is best not to be drawn into conversations with the press or the public, and I have found that a non-commital "*I have been sent to find out what happened so as to prevent it happening somewhere else*" is about the extent of what one needs to say if challenged.

Sometimes it is necessary to take a party of visitors or students to the site of a disaster for educational or training purposes. In my experience it is sadly necessary to give a briefing on the appropriate behaviour as well as the safety briefing, especially with students, as there may still be people at the location who have been affected and remain sensitive ever after, and may be upset by inappropriate demeanour.

NOT A DISASTER, BUT JUST AN OBVIOUS LANDSLIDE

A much more common scenario, thankfully, is to make a site visit to an obvious landslide, which will be very slow moving or stationary. In effect, it has stopped being an *event*, and instead has become a *landscape feature*. It is always easier to deal with the inanimate than with people. This part of an investigation is sometimes referred to as a 'walkover survey'. It is a term I dislike greatly, not least because it is often mistaken for something casual or even frivolous. On the other hand, 'geomorphological mapping' is something more specific and needs to be carried out by a geomorphologist! I generally prefer 'site inspection' or 'detailed site inspection'.

During such an inspection it is necessary to record your observations. The best place to do this is on a large-scale map. We are fortunate in the UK that the Ordnance Survey has been preparing large-scale maps since the middle of the 19th century, and copies are readily available, especially of the current map. I'll discuss maps when it comes time to describe the 'desk study', another term that I dislike, preferring 'review of documents'. If a map is not available, then an air photograph might do, including a print of one of the Internet mapping systems like Google Earth. Alternatively, one can prepare a map oneself or with the assistance of surveyors. Air photographs can be taken from drones as well as from aircraft, with photogrammetric maps prepared from either. Handheld GPS is good if the landslide is large enough so that more than metre scale positional inaccuracy is not too damaging, but even then, the level information is rarely good enough. The great advantage of preparing a map oneself is that you record the features at the same time using the same basic system.

The features that are most valuable to identify are those that tell us about the size or extent of the landslide. Some features tell us the kind of landslide that has to be dealt with, and others even how deep it probably is. Other features that can sometimes be seen may also give clues as to why the landslide occurred in the first place, or why it is active. It's a matter of whether the features are present, and whether they can be seen and recognised. They also need to be recorded.

NOT OBVIOUSLY A LANDSLIDE SITE

There are various reasons why a landslide might not *look* like a landslide. It may have been bulldozed or ploughed, so the features are very subtle or even non-existent, but far more likely is that the surveyor doesn't know what he[16] is looking for, or can't work out the significance of what he is seeing, and sometimes, the scale is so large that he doesn't realise that the landscape contains a huge landslide. Alternatively, he sees the shallow and superficial landslides, but misses the fact that they conceal something on an altogether larger scale underneath. We are all conditioned by our experience here, and like the old adage that the best geologist is the one who has seen the most rocks, you need to see lots of examples before you can recognise them when you see them. If you don't have the opportunity to see them in the field for yourself, then this is a case where the internet,[17] or your library, is your friend.

It's also difficult to recognise some landslides that have been partly or wholly stabilised, for example in infrastructure earthworks.

Add to those difficulties the fact that some landslides look like something else, or those other things, like a disused quarry or a corrie in an upland areas may look a lot like a landslide, and you have the recipe for either missing something or going off on a wild goose chase.

If obvious landslide features cannot be seen, it certainly does not prove that there are no landslide problems. The absence of evidence cannot be assumed to be evidence of absence. Evidence can be obtained by looking further afield, in sites with the same geology or in the same geographical context. Sometimes the evidence is rather close at hand – in the next field or building plot. The evidence is sometimes from nearby completed stabilisation schemes or in archival materials. A classic case is an abandoned cliff: a coastal or riverine slope defended naturally from erosion by the growth of a beach, spit, saltmarsh, *etc*., on the coast, or where the course of a river meanders away and leaves the river cliff to degrade into a slope containing old, degraded, and difficult-to-recognise landslides.

Essentially, if the geology and geography are right for landsliding and you can't find the telltale signs, the question must be asked "*why isn't there any sign of it?*" At that point, it is probably time to call in someone with more experience to look, and then to learn from them.

WHAT TO LOOK FOR – THE EXTENT OF THE LANDSLIDE

Features that delineate the extent of landslides include toe bulges and head scarps. The appearance of toe bulges depends on the inclination at which the toe breaks out. Where the mode of failure involves base failure, which is where the slip surface emerges beyond the toe of the slope – where *beyond* is from the perspective of an observer standing at the top of the slope and looking out over the landslide – then the slip surface at the base of the landslide inevitably rises as it emerges, the toe region of the slide may be compressed as well, and the feature that is seen is a big hummock. A landslide may have multiple toes even in a single event, as the thrust and loading may cause the soil surface even further away (again from the same perspective) to buckle, and especially where the soil is the result of previous landslides, it may shear out along pre-existing slip surfaces. Where landslides occur on the coast or into river channels, the toe breakout position may not be entirely obvious at high tide or when the river is in spate.

Shallow landslides tend to override the ground at the toe, and occasionally there are multiple toes, each representing an episode of movement, overlaid on each other. In grassland, the toe may appear like the fold in fabric with the turf overturned and acting to retain soil. In some cases, it is possible to probe under the overturned turf. Trees that are pushed by a landslide at its toe may retain the toe or be pushed over, and this also applies to retaining structures and buildings, with possible collapse of walls on the loaded side. Shallow landslides also tend to follow the terrain slope, and sometimes change direction as they do so. Toe features are normally convex in plan in the direction of movement.

At the head of a landslide, the rear (or head) scarp face is often apparent as a step. These steps can be small in the case of shallow landslides, where the slip surface is at no great depth, or high where the landslide is deep seated. The ground surface above the actual head scarp may show tension cracks, which are an indication of overstress but not yet to the point of including more soil in the landslide. If the slide continues to move, the tension cracks delineate the successive failures that may take place. In a really recent landslide in clays the exposed slip surface may still exhibit the polished and slickensided appearance of clay that has been sheared. Head scarps of many landslides are often convex in plan in the upslope direction, but compound landslides may have long straight sections of rear scarp if the landslide is comparatively wide. But do beware, small head scarps may appear, not because of a failure, but due to differential settlement in fill, especially adjacent to a structure.

Finally, at the sides of a landslide, lateral shears may be seen. Occasionally, lateral ridges formed from soil that has spilled over onto adjacent unslipped ground may be generated. These are sometimes called levees.[18]

When a landslide has aged, being subject to the effects of weather, and walked over by animals or people, the features gradually lose their freshness and become blurred. They therefore become more and more subtle as a landslide ages, and correspondingly more difficult to see, even if there has been no deliberate attempt to obliterate them.

WHAT TO LOOK FOR – THE NATURE OF THE LANDSLIDE

Features that reveal the *nature* of a landslide are usually within the footprint, identified by the toe, head scarp, and lateral features. One of them is a *graben*, which is a sort of depression at the head of a landslide, bounded on one side by the rear scarp and on the other, by a counter- or antithetic (opposite sense) scarp. Grabens usually occur in landslides that have a change of curvature or orientation is the shape of the slip surface, requiring internal deformations that propagate to the surface. They are therefore common in compound landslides that have a structurally controlled section that may be approximately plane, together with a section rising up to the head scarp that is curved. Typically, the maximum depth to the basal slip surface in the graben region is of the same magnitude as the width of the graben,[19] which is a guide, not a guarantee, but is useful when planning a ground investigation with boreholes.

Grabens also appear with shallow, translational, landslides, but they widen from their original width if the landslide continues to move, and then they are more extensional rather than subsidence features, and the above loose correlation doesn't apply. You can visualise this best when a landslide takes the form of strata dipping (and sliding) at the same angle as the terrain slope – a trough at the head of the landslide would not then constitute a graben.

An analogous feature in some compound landslides instead of a graben is a backtilted block. We often see this when the top of the slope is occupied by a caprock which subsides and rotates. The geometry of the slip surface is similar to that creating a graben, and so the correlation between width and depth is also similar.

In structural geology, grabens alternate with ridges called 'horsts' where the landscape is subject to extension. Landslides that spread a lot may exhibit a series of grabens and horsts in much the same way. Cases of spreading reported in the literature include pathologically large runout slides, and some materials that degenerate into flows because the loss of strength on shearing is extreme. Neither tends to be investigated fully or well, in the former case because of scale and in the latter case because of access difficulties.

Some landslides also show multiple scarps or lines of cracks. The lines of cracks usually indicate changes of gradient or direction in the underlying slip surface, although they can simply be scarps in course of development. Multiple scarps occur for a variety of reasons, including the toe region becoming unstable independently of the main slide.[20] This situation is particularly the case with landslides perched up in a slope where the toe region falls off as the slide moves forward. Scarps may also occur when the slip surface steps down from one weak bed to another. If the vertical spacing of low-inclination weak beds in a slope is large, then the slope develops into a very obvious 'staircase' with a separate landslide on each step, but if the vertical spacing is smaller or the landslides are particularly well-developed, the individual steps coalesce, and they are then obscured except, perhaps, for a scarp associated with each.

Landslides associated with fills over extremely weak sediments are often associated with extrusion of soft, remoulded, material which emerges at the surface through tension cracks and intermediate slide scarps and might just be an application for the description *lateral spreading*.

The evidence for past rockfalls lies in the existence on site of scree or talus slopes: accumulations of debris at around 40° inclination formed from piles of rock fragments. Rocks with larger run-out are often cleared up, but the scree slopes are left. *Scree* is an old English word, but *talus* simply means slope (or 'bank'), so a talus slope is a slope slope,[21] just in case that wasn't clear when using just a single word.

SOIL AND ROCK PROPERTIES

The relevant strength parameters for landslides in soil and rock obey similar rules, being dependent on effective stresses and having both a cohesive component and an angle of shearing resistance, denoted by the symbols c' and ϕ', with the cohesion using the lowercase Roman letter c and the angle of shearing resistance the Greek letter ϕ, and with both having the $'$ mark to denote that they are determined with respect to effective stress. The 'frictional' strength component is the product of the effective normal stress σ' and the tangent of the angle of normal stress, *i.e.* $\sigma' \tan \phi'$.

Rocks usually exhibit large and significant cohesion and a high angle of effective shearing resistance, and so landsliding in rock is most commonly a function of the discontinuities in the rock mass rather than failure of the solid material between those discontinuities. The exception is the case of mountain rockslides, where the scale of the landslide is such that the stress levels can overcome the intact strength of the rock. Even then, the bulk properties are influenced to a degree by the discontinuities, and for this, relationships such as the Hoek-Brown method[22] are used to assign soil-like parameters to the rock mass overall. However, as most discontinuities exhibit small, negligible or zero cohesion, many small-scale instabilities are dominated by the angle of shearing resistance of the discontinuities, and this may well be less than the angle of shearing resistance of the intact material, especially if the discontinuity has been sheared and therefore is likely not to have significant cohesion. Caution is therefore required when using the Hoek-Brown parameters where instability can develop along major discontinuities such as faults, old rockslide slip surfaces, *etc.*

Intact rock parameters are complex, and depend on rock type, mineralogy, type and grade of metamorphism, and many other factors.

In soils, the cohesion element is commonly a function of overconsolidation, especially in clays, accentuated by the presence of minerals such as calcite and iron compounds that are cementitious. Cohesion is again a function of discontinuities such as fissures in overconsolidated soils and clay shales, so a safe assumption is that there is only small or zero cohesion in such soils. In granular soils, the angle of shearing resistance is somewhat dependent on soil density and the angularity of particles, but usually large compared to cohesive soils, where the angle of shearing resistance depends greatly on the clay content and clay mineralogy. Landslide slip surfaces always display negligible cohesion when freshly formed, but the presence of cementitious agents and other processes may give a slight regain of strength in landslides that are not moving. The angle of shearing resistance in all materials is lower after shearing than when undisturbed, but in some clays the drop in this parameter may be dramatic: most in smectites, and least in kaolinites compared to the intermediate behaviour of illites, which form the majority of clay strata deposited in saline environments. Tropical weathering creates halloysites, which may prove to have characteristics somewhere between illites and smectites.

The sheared strength of soils is termed the *residual strength*, and is denoted in symbols for strength parameters with a subscript lowercase letter

r, although the cohesion is small or zero, and the residual angle of shearing resistance is less than the peak angle in intact soil.

The use of cohesion and an angle of shearing resistance implies that the soil has strength at zero effective stress, and also that strength increases linearly with increasing effective stress. This is approximately true over small stress ranges, and where the relationship is nonlinear, it is better dealt with in any form of analysis by subdividing the materials into different zones rather than by increasing the sophistication of the constitutive relationship for soil strength.

Cohesive soils are so named not necessarily because they exhibit large cohesion, but because they do not fall apart into particles in a hand specimen. Such soils can sometimes also be remoulded by hand if they are weak enough. Cohesive soils almost inevitably contain clay minerals, and the particles cohere or stick to each other or adhere to other things by processes involving mineral bonds facilitated by the water content and by the pressures in the pore water, especially when the soil is unsaturated. This gives rise to something called the *undrained strength*, or the Roman lowercase letter s subscripted with u for undrained. Soils that are saturated or nearly so exhibit only a cohesive component to the undrained strength, so s_u is sometimes written c_u, but in soils with intermediate levels of saturation, there may be an undrained angle of shearing resistance ϕ_u as well. Soils exhibiting undrained strength often lose part of it when remoulded, a phenomenon known as *sensitivity*, but when all or almost all strength is lost, we call that *liquefaction*.

Laboratory test procedures exist to determine all the relevant properties of both soil and rock, but there is always a question as to the relevance of the result on a particular test specimen to the deposit as a whole.

ON THE BEACH – SOIL MECHANICS FOR CHILDREN

Some of the important lessons in soil mechanics can literally be seen as child's play. For example, if you make a heap of dry sand on a plank, and gradually lift up one end, eventually you generate a flow. Then, if you put a row of flat stones or tiles on edge on the plank after cleaning off the sand and returning the plank to the horizontal, then tilt it again, you will get a topple. Now repeat the exercise with the stones stacked flat, and you get a slide.[23]

Some other aspects of soil behaviour are easily demonstrated on a sandy beach as the tide goes out. The sand can be made to give up some of its water content if you 'paddle' it with a foot, but applying pressure to the surface (and therefore shear, under the effect of which the soil dilates), the water is sucked back in. This is a field demonstration of the effects of the pore pressure parameters A and B!

When building sandcastles, you need damp sand. Then, the multiple capillary menisci ensure that the sand holds together, demonstrating soil suction. This suction is lost when the tide comes in and, even if we discount the effects of waves, destroys the suction. Suction can also be lost if the sun is intense and dries the sand, making your elaborate sandcastle turrets and walls collapse into conical dry sand mounds. If the sand is dirty, for example, containing clay or some minerals, the shapes may be retained when the sand dries out, and in that case, you have demonstrated cohesion, which in many cases is simply the result of surface tension in the water and therefore a state of 'suction' as the air is at atmospheric pressure, and the water pressures must be less. Incidentally, those conical mounds stand at an angle equal to the angle of shearing resistance for that kind of loose sand, thus giving rise to the deprecated alternative name 'angle of repose'.

Budding dam engineers in their infancy may also like to dam small streams crossing sandy beaches. They soon discover the damage wreaked by overtopping, or if they can keep pace with the rising water levels, then they will find out about seepage or *internal* erosion.

In my experience, however, few children are ready for the lecture, even though they spent the day building sandcastles that the incoming tide ruined![24]

NOTES

1 We also have 'Lost *in* a landslide, no escape from reality' in 'Bohemian Rhapsody' by the band Queen. Fleetwood Mac seem to have got the idea of a mountain landslide in one of their tracks, although Olivia Newton-John's landslide seems to me more like something that would be inflicted by Hannibal Lecter!

2 Terzaghi classified slip circles as those that emerged exactly at the slope toe as a 'toe circle', those that emerged beyond the toe of the slope as a 'base failure' and those that emerged in the slope as a 'slope failure', so perhaps you will see why I am not keen on the use of this phrase to cover landslides of all sorts.

3 Skempton appears to have consulted on the Muirhead Dam in Scotland, which was being built in the 1940s. It reached a certain height, but could not be built any taller, as fill added to the top simply made the dam wider. Later, the loss of strength on shearing was given the name 'brittleness'. Skempton taught me in my Imperial College MSc that compacted clays had no brittleness. He repeated this in his 70th birthday lecture at the ICE on the day that the Carsington Dam failed. I had known for many years that he was wrong, and it didn't make it into the final print version of his lecture. He later graciously conceded that it had been wrong. No doubt there was some sensitivity about the problems at Muirhead, because in his 1962 book *Dam Geology* (Butterworths), R.C.S. Walters refers to it several times as 'Muirfoot'!

4 As you might find, for example, in Hutchinson's 1988 paper: Hutchinson, J. N. (1988) General report: Morphological and geotechnical parameters of landslides in relation to geology and hydrogeology. *Landslides: Proceedings of the 1988 ISL, Lausanne. Balkema, Rotterdam*, 1, 3–55.

5 Cruden, D. M. & Varnes, D. J. Landslides, types and processes. Chapter 3 in A. K. Turner & R. L. Schuster (Editors) *Landslides Investigation and Mitigation, Special Report 247*. National Research Council (US) Transportation Research Board, 36–75. There were 2 previous books on the general topic, the first published in 1957 (Report 29) and the second in 1978 (Report 176).

6 Dikau, R., Brunsden, D., Schrott, L. & Ibsen, M-L. (Editors) (1996) *Landslide Recognition: Identification, Movement and Causes*. Wiley, International Association of Geomorphologists Publication No. 5 and Report 1 of the European Commission Environment Programme Contract No EV5V CT94–0454. The book is difficult to find now, as it was produced in quite a small print run, but if you are very, very, lucky, you might find one of the first print run where the title on the cover was misspelled with 'Courses' instead of 'Causes'!

7 Eisbacher, G. H. & Clague, J. (1984) *Destructive Mass Movements in High Mountains: Hazard and Management*. Geological Survey of Canada, 84–16. This book is also out of print. Fortunately, a copy was scanned and is on the website of the Canadian Geological Survey.

8 Quick clays are marine deposits that are above sea level because of tectonic uplift. Rain leaches out the salt content and leaves the structure of particles in a state where some disturbance makes the whole structure collapse or liquefy. The word 'quick' is used in its sense of 'alive' as in 'The quick and the dead', and not relating to the rate with which collapse can propagate through that part of the clay deposit which is quick, the speed of which is terrifying.

9 It is a well-known phenomenon that synclines run through many mountains and smaller hills. This is sometimes called 'relief inversion' or 'inversion of relief'.

10 There is an extensive literature on the landslide from a colliery spoil tip that killed 144 people in the UK's worst-ever landslide disaster. A review by Bishop covers the problem in some depth. (Bishop, A. W. (1972) The stability of tips and spoil heaps. *Quarterly Journal of Engineering Geology and Hydrogeology*, 6, 335–376).

11 This note requires the citation of 3 papers: Hutchinson, J. N. (1970) A coastal mudflow on the London Clay cliffs at Beltinge, North Kent. *Géotechnique, 20* (4), 412–438; Hutchinson, J. N. & Bhandari, R. K. (1971) Undrained loading. A fundamental mechanism of mudflows and other mass movements. *Géotechnique, 21* (4), 412–438; Hutchinson, J. N., Prior, D. B. & Stephens, N. (1974) Potentially dangerous surges in an Antrim mudslide. *Quarterly Journal of Engineering Geology and Hydrogeology*, 7 (4), 363–376.

12 Hutchinson and I introduced this term in a paper: Hutchinson, J. N. & Bromhead, E. N. (2002) Keynote paper: Isle of Wight landslides. *Conference on Instability: Planning & Management. Thomas Telford, May 2002*, 1–37. It transpires that SPHs are often smectite enriched thin strata.

13 What about: Evans, S. G. & DeGraff, J. V. (2002) *Catastrophic Landslides: Effects, Occurrence, and Mechanisms*. Geological Society of America. Reviews in Engineering Geology Volume XV.

14 There wasn't at the time. However, a conference accepted a paper on the subject of turfy soil. You may guess that the editor for the conference proceedings was neither a native English speaker nor an expert on peat.

15 Or coffee in the US, where the delights of tea made in an enamelled teapot and served with whatever variety of milk, preferably 'condensed' and therefore presweetened, and drunk from a grubby site mug, are unknown.

16 'He' generally includes 'she' and every subdivision in between. English is a funny language at times. I do hope that no-one is offended by this.
17 At the time of writing, Professor David Petley maintains a 'Landslide Blog' on the AGU website that contains reports of landslides from all over the world, many of which include videos taken by bystanders on site.
18 A horrible word with multiple meanings.
19 Cruden, D. M., Thomson, S. & Hoffmann, B. A. (1991) Observation of graben geometry in landslides. In R. J. Chandler (Editor) *Slope Stability Engineering – Developments and Applications*. Thomas Telford, 33–35. In conversation with Cruden, I discovered that he had subsequently changed his mind, but I have nevertheless found a rough equivalence to be of value.
20 Hutchinson, in his 1969 paper in Géotechnique on the Folkestone Warren landslides, classified the main or massive landslides as type M, and the coastal margin slides as rotational or type R. They share the same, bedding-controlled, basal slip surface. Hutchinson, J. N. (1969) A reconsideration of the Folkestone Warren landslides. *Géotechnique*, *18* (1), 6–38. It was subsequently discovered that while the M-type landslides had been by a system of drainage adits and a toe weighting berm, the R-type landslides continued to move, despite having their Factor of Safety improved by the berm much more than the M-type landslides. A reason for this could be leakage from the drainage adits that were deformed and fractured as they crossed into the R-type slides and consequently leaked.
21 Just like Lake Windermere is *Lake Winderlake*, and the River Avon is *River River*. Sometimes the origins of English words are far from clear.
22 Hoek, E. & Brown, E. T. (1997) Practical estimates of rock mass strength. *International Journal of Rock Mechanics and Mining Sciences*, *34* (8), 1165–1186. Note that the paper says it is *mass* strength, and also that it is an *estimate*.
23 During COVID lockdown, you could do the same experiment indoors with books. I did.
24 Kids in the car also complain if you point out too many landslides on the road! Mine have grown up, and now prefer to be the driver. They still prefer me to keep my geotechnical observations to myself.

Chapter 2

Ground investigations

PLANNING AND SEQUENCING

The purpose of a *ground investigation* as a constituent part of investigating and stabilising a slide or flow landslide is to determine the nature of the materials in the ground and how they are distributed. A large part of this is to determine where the slip surface (if there is one) is located,[1] and an equally large part is associated with investigating the groundwater conditions. In the case of falls, it is a matter of inspecting the source area to determine not just whether or not future falls are likely, as they most probably are, but what must be done to stop new falls from happening – and that is a question of surveying the discontinuities more than anything else.

After about a half-century in the business, including sharing my teaching with a number of very competent geotechnical engineers, I have come to the general conclusion that there are at least two types of approach to design with two very different kinds of geotechnical engineers involved.[2] The first kind is the engineer whose interests lie in foundations. These individuals generally have comparatively little interest in the finer points of geology, and are interested in ground investigations as a means of finding materials that are strong enough to bear the foundation loads without much settlement, and they are therefore content to have widely spaced samples, drillers' logs, and boreholes distributed across a site so as to cover the area. The other kind, of which I have reluctantly come to accept that I am, is interested in the weaker parts of the succession that are rarely identified well in drillers' logs, and requires a particular plan arrangement of boreholes. The former is often content with the level of geology taught in undergraduate civil engineering courses, but the latter knows that it was insufficient. This probably means that the former retains friendships with his student colleagues who went off to become structural engineers, whereas the latter nurtures friendships with all manner of geologists[3] and geomorphologists or, indeed, comes from one of those backgrounds! Of course, I am not entirely serious in these

DOI: 10.1201/9781003428169-2

remarks, but like many things said in jest, there is an element of underlying truth in them.

As well as that qualitative difference, there are, of course, other differences between engineers with geotechnical backgrounds and postgraduate experience or education, and those without. I find that the most important partition occurs with an understanding or not of the pore water pressure behaviour that leads to undrained strength. There are probably other equally important 'grade boundaries'.

When investigating a landslide with a recognisable surface expression, it is not sufficient to put boreholes solely in the footprint of the landslide, it is necessary to also put them in unslipped ground, usually beyond the head of the landslide, in order to establish the stratigraphy relevant to the site. In some geologies, especially sedimentary geology with relatively thick beds, it may be possible in your ground model to project the beds from the solid through the landslide, and thus gain insight into what has slipped and what has not. This approach may not work well – or at all – if the geology is faulted, folded or smashed up in a mountain belt. Boreholes beyond the toe may be helpful if the landslide toe has encroached on coastal sediments or other weak materials, not least because they will reveal whether or not those materials can support a toe berm as an element in any stabilisation scheme.

After you have established the context of a landslide in this way, you can start on the boreholes *through* the landslide. Sometimes, landslides are sufficiently small to be investigated by a few boreholes, but the larger and more complex the landslides are, the more boreholes are called for. My preferred approach is to identify the main section that you will investigate, termed the *principal cross section*. Then, with boreholes arranged along it, the job of drawing up a ground model is simplified because there is no doubt what is on the section and what is not. If boreholes are initially scattered over the site of the landslide, there will always be uncertainty about how they should be projected onto the principal cross section (that is, if projection is meaningful at all). Two choices that are obvious are, firstly, to simply project at right angles to the section, and the other is to project along the strike, if a consistent dip across the site has been found. Neither is entirely satisfactory, and having the boreholes on a single section is far better for the development of a ground model.

Once a ground model on the principal cross section has been developed, then investigations on other sections, including those that cross the landslide transversely, may be undertaken. The later sections may reveal things that were missed or ignored when developing the principal cross section.

No matter how carefully you *plan* the layout of the investigatory boreholes, you must be prepared to revise everything in the light of findings of the early phases, and again and again as more data comes in. Accordingly, I favour site logging as soon as the samples or cores are available, followed by plotting up the data on a drawing while on site.

RECORDS OF PAST INVESTIGATIONS

Several times in my career, I have been faced with a site that has had numerous ground investigations spanning several decades, and in the worst cases, the records were incomplete. I have lost count of the reports, complete with borehole records and test results that were missing the site plan, taken from the acetate pocket at the back, and never replaced, so that now it is lost. Or the corresponding case, the disembodied site plan, detached from the report which contains the borehole records, and which is now missing. Those site plans from the past were usually dyeline prints, fading through exposure to sunlight, and not fixed to anything definite, but with levels related to some arbitrary benchmark on a feature that was removed decades ago. Inevitably, site plans tell you 'Do not scale', but there is little choice, particularly as the scale of the older drawings is inevitably pre-metric. A great advantage of the modern survey is that a borehole record is likely to have GPS coordinates and levels, and that makes the report self-contained even without a site plan.

Developments in technology, however, do little to help in interpreting incomplete historic investigations, but they do provide a lesson in what really needs to be done to preserve contemporary investigations for future use. At least a print version will continue to be readable if it survives physically, but a digital version also has the added problem that even if a file exists, its format may be obsolete, and then the file will simply be unreadable.

As an example of the difficulties that a too-rapid evolution in datafile formats can cause, we can take the results of the Selborne Controlled Slope Failure Experiment. In order to disseminate the results, they were collated onto a CD. We used the then most popular word processor format, which at the time was WordPerfect, and a popular spreadsheet, Quattro Pro. The spreadsheets of results included various graphs which could be examined to see directly where the results had come from. However, later versions of Quattro Pro had compatibility issues with the files made at the time, and when Excel became more popular, it could import the Quattro Pro files, but definitely lost all the graphics. The latest version of Excel doesn't even have the import filter. Similar issues occurred with WordPerfect-to-Word conversion; with the photographs originally in Kodak PhotoCD format and later in JPG; and with a collection of presentation slides drawn in an early version of CorelDRAW no longer readable by the current version. What is more, with many computers now no longer sporting CD drives, and CDs known to lose data with age, it has been an uphill struggle to maintain the valuable dataset for use by researchers.

No doubt ground investigations stored in electronic form have been lost for similar reasons of software incompatibility and hardware evolution.

There is also the problem of a landslide site appearing in a paper whose authors are deceased, and no other trace of their work exists. Almost certainly the source materials are lost. It is simply a question of getting on with it and starting again from scratch.

Often, a paper is produced that is simply a reworking of older information. It is sometimes difficult to discover if the new interpretation really contains new observations as well as just a reinterpretation, and if possible, you should read the original sources as well as the most recent paper. Sometimes there were mistakes in this revision and republication process, and sometimes the original work is not even cited properly, and is only discoverable from anecdotal evidence.

THE TRIAL PIT, TRENCH, OR CLIFF FACE

Early in my career, a colleague was killed in a trial pit collapse, and I know of other cases. It made me nervous of entering trial pits to do logging, and I am relieved that nowadays health and safety legislation in the UK thankfully forbids such unsafe practices.

Even if you cannot enter a pit, that does not mean that trial excavations have no role to play in a ground investigation, as the more obvious elements in the soil succession can still be logged at a distance. One of the best positions to see what is down the hole is the excavator driver's seat, and most geotechnical engineers will not be driving the excavator themselves. It is sensible therefore to integrate the driver into the team on site by giving him a useful, friendly, induction into what is required, and how you want to have the trial pits excavated, what signals you mean to use to get him to stop so that you can approach the pit or spoil heap, and so on. I have found many excavator drivers take an interest in what they are digging through and are delighted to be helpful in identifying the material changes seen from the cab or felt through the controls of the excavator.

Sometimes, the alignment of a pit is chosen to suit access or positioning of the excavator, for example at right angles to a toe feature, with the excavator positioned on unslipped ground, but where possible, arrange it so that sunlight enters the pit, and it is not in the shadow of the excavator. Light can also enter the pit if it is not a slit trench but is opened out on the side that would in any case be in shadow. It is handy to use a laser (or LED) pointer to indicate features to a colleague, but beware if flying to a foreign destination, as these items are forbidden[4] in cabin baggage. I cover this, and other parts of my recommended toolkit, in Chapter 10.

Trial pits can only penetrate to relatively shallow depths, and are at their most useful with relatively shallow landslides that are only a few metres deep. In larger landslides they are useful in the region of the toe, and it is here that a slip surface is most readily detected. It is therefore a good location to start with, because it enables you to train yourself and colleagues[5] (including the excavator driver) in what to look for on the particular site. In clay soils the excavator bucket shears and smears the clay, but slip surfaces open up as the bucket pulls the soil away, and therefore, the difference, once seen, can be remembered.

Some of the techniques developed for finding slip surfaces in trial pits are useful if one has a trench that is adequately supported to allow entry and allows inspection of the soil sequence. This may be the case where support has been provided to allow pipe joints to be made and inspected. The techniques are also applicable when inspecting a cliff section.

Essentially, the technique of finding slip surfaces is to clean the face by removing smeared soils or slope wash (on an exposed face) and then identify the difference between the textures of the landslide debris above and intact material below. The characteristic textures of the soil inside the landslide will be variable and depend in a large part on what soils are involved. Landslides that entrain artefacts in the debris, such as tin cans, plastics, wood, and so on are obviously easy to recognise, and so are inclusions of natural materials from higher up the slope. Block slides along bedding are more difficult, as the correct stratigraphic sequence is preserved, but the slide mass is almost always subject to more deformations than the parent body underneath, which causes the soil to have more discontinuities such as the fissures in stiff clays, and this allows you to recognise them.

Mudslide (or mudflow) fabrics when exposed have a texture where the air inclusions look like pinholes, and other fabrics are blocky. Once again, this is something you learn from experience, from personal tuition in the field, or in the sample description shed.

Once the two areas are recognised, you pick your way up or down using a stiff-bladed knife to prise out small pieces of the soil from the face. Eventually, a piece will break off along the slip surface. Further small excavation will confirm it, as the slip surface in clays has a polished and striated or fluted appearance, with much less striation if it has developed along bedding.

Occasionally, the slip surface may exhibit a different colouration due to the movement of water along it, so that for instance, in soils containing iron compounds that weather to brown, water may carry organic content that reduces the slip surface to a light-blue colour.

In a rock slope, of course, a slip surface will be indistinguishable from a fault, for the very obvious reason that they are essentially the same thing!

BOREHOLES

There are at least four technical elements to using boreholes to investigate landslides. The first is what equipment you use to make the hole; the second is how you take samples; the third is what in situ tests you might do, and finally, there is a matter of what instruments are left in place down the borehole. Those are the choices to make regarding the individual boreholes, but there is a further set of choices to make that include how many boreholes to drill, and what layout or arrangement in plan is the best for your purpose, as described above.

The preferred method of drilling varies from country to country, and usually depends on the rock and soil types most commonly encountered, and therefore the most common equipment held in ground investigation contractors' inventories. In the UK, and especially until comparatively recently, the dominant drilling equipment was that for percussive boring, although for small-scale and relatively short boreholes, many contractors use the man-portable driven tube sampler with or without a window. Percussive drilling is also known colloquially as 'shell and auger', with samples taken using 100mm nominal diameter open drive samplers about 450mm long in a borehole with 200mm or occasionally 150mm diameter. I personally favour continuous flight auger drilling for instrumentation holes, but those rigs are not well suited to sampling.

In other countries, various other methods such as wash boring and rotary drilling are favoured.

I do not think any of the techniques with which I am familiar[6] are very suitable for landslide investigation. Open drive samples are too short, and the technique often damages the sample. Most rotary coring is too narrow to split the cores to look for changes in soil texture. But the techniques are all we have, so it's Hobson's Choice.

Particular techniques that I find to be in the way of doing the job properly are in the case of percussive boring to only take samples say every metre, which basically means that over 50% of the soil profile isn't logged at all. Wider spacing is worse. Then, poor recovery makes things even less useful. In rotary coring, the way that the core is extracted from the core barrel damages it – I have seen core blown out of a core barrel by water pressure and fired off like cannon balls. Too-short core boxes where the driller breaks the core runs up to fit may help portability, but really do not help the logger. Then there are cores left outside for days in climatic extremes before they are logged. And before the list gets too long, rotary coring with the wrong flush medium is unhelpful. Many of the techniques I deprecate are perfectly adequate for the needs of the foundations specialist – perhaps – but they are not good for the landslide investigator.

LOGGING

The principles of soil and rock description are well understood and practised daily by engineering geologists and geotechnical engineers. It is a basic skill best learnt on the job, although it is described in a number of useful publications[7] to help with both the basics and the finer points. The only additional skill needed in landslide investigation is to be able to recognise a slip surface – if there is one to be seen in whatever cores or samples there are.

An important factor in the preparation of borehole logs is that they should truly reflect what is observed in the samples or core. I have several times seen some obvious and non-obvious mistakes in this. In one case, the logging was done skilfully and professionally in the geologists' notebooks. Periodically, they would break off from the core boxes and move to an office, where they wrote out their logs on blank forms, combining their observations with the drillers' records. The forms were sent to a helper who typed the information up in software that prepared the final logs. The check was to compare the final logs with the handwritten drafts. Note, not with the geologists' notebooks or even the cores. The project was never built, perhaps fortunately.

Another failed check I experienced was for the logging to be done in a foreign language, then translated by the chief of the ground investigation organisation, whom no-one dared question. Was it a good translation? Who can tell?

In my view, logging should be done on site, and soon after the samples have been taken, with the geologist's log compared to the driller's log at the first possible opportunity, and disparities checked and resolved, returning to the original records at every stage in the process as a fundamental principle of quality assurance. In any case of doubt, further boreholes are required, and in my view, the site should not be closed until a satisfactory ground model has been developed.

THE GROUND MODEL

In the early days of my career, I always had access to a full-size draughtsman's drawing board, and would develop ground models on paper of about A1 size, although at first Imperial sizes were used. Eventually, I discovered that although the large size was useful for drawing, ground models on larger than A3 size were not really necessary, because if you could not detect any geometric difference at a scale that would fit, then it wouldn't show up in the results of a stability analysis in a significant way. I prefer to use paper that is already printed with graph squares, and having drawn a cross section outline, I ink it in and then proceed in pencil with the subsurface details. In Chapter 10, I describe my 'office' toolkit for doing drawings and analyses.

Drawing in ink is now becoming something of a lost art, and my younger colleagues often prefer to work from scratch in a CAD package.

The starting point in the development of a ground model is to begin with the topography, then to mark up the solid geology, and then finally to add the boreholes one by one, entering the stratigraphy and logged slip surface positions, the latter being joined up progressively.

A good hint that you are drawing details above the slip surface is if the stratigraphy is displaced from where it might have been or is indicated to have been by the boreholes off the landslide footprint. Sadly, if the differentiation of the strata cannot be done visually, but needs palaeontological or micropalaeontological studies, the results won't come in until long after the fieldwork has been completed.

Some of the mistakes I have seen have been appalling, such as ignoring the logged slip surfaces to fit an idealised slip circle, or to draw in slip surfaces to meet some prejudice about where the slip surface *should* lie,[8] regardless of what was logged or indicated in instruments.

INSTRUMENTATION AND MONITORING

Geotechnical instrumentation is available for a variety of tasks, but in landslide engineering the two main types of downhole instrumentation measure groundwater pressures (piezometers) or lateral ground deformations (inclinometers) with a third measurement – of settlement – available but uncommon. There are many types of each.[9] Other types of instrument can measure in-situ earth pressures, with typical applications in retaining wall monitoring.

I have a rule of thumb that I call my 1:1:1 principle (one-one-one principle). That is to budget equal amounts for purchase and installation, reading and interpreting. Far too many instruments are installed and, at best, read a few times during the rest of the ground investigation and then only once or twice afterwards, if that. The 1:1:1 principle recognises that as far as reading is concerned, there is a range of options about how it is done. At one end of the scale there is manual reading, but this requires somebody to visit the site periodically and lower something down the borehole. In the case of tube piezometers, this is just a sensor that makes a sound when it enters the water and closes a circuit. These instruments are normally provided with a calibrated cable attached to a reel, but if used on a muddy site, the cable becomes dirty, and that can contaminate the tube, leading to false readings.

Manual readings from geotechnical instrumentation are far more expensive than many people realise. There is a question of the salary and employment costs of the technician who does the readings and their expenses: travel costs, subsistence, and so on. Travel costs are very obvious if air travel is involved, but even mileage claims or the cost of use of company vehicles needs to be taken into account.

Many types of piezometers need elaborate read-outs, which must be maintained and calibrated periodically. The best regime of calibration is prior to every use, and even though modern electronics are far more stable than they used to be, the question must be how far back should you go before you can trust the readings if the calibration has drifted? The answer is only that you have to go back to the previous time the readout was calibrated.

The problem of calibration is exacerbated if the readings are taken automatically. Some systems record the data periodically on a datalogger, which needs visits to site to download the data. Other systems transmit the data back to base *via* cellular telephony or even a direct satellite link. Such systems are expensive to purchase and, in many locations, are at risk from theft or vandalism. Instrumentation installations are very obviously valuable, more so if they are protected with fencing, and this attracts thieves and vandals. In some countries, solar panels provide enough charge to power the electronics, but elsewhere they may run out of charge in the winter or in long periods of bad weather. Sometimes, the equipment freezes up. Regular maintenance visits are therefore essential, and they are expensive too.

The situation is even worse where an inclinometer probe is used, as dirt can foul the keyways in the access tube. The probe is much heavier than that on a piezometer 'dipmeter' and needs a correspondingly heavy cable, so for a deep installation it may prove impossible to carry the instrument to site, and vehicular access is needed.

So the reading is obviously expensive, but many people are astonished that its costs can mount up to match the costs of purchase and installation. I am not. However, it is sometimes difficult to understand why the costs of interpretation can be so high. The problem here is that the interpretation must be done by experts, and experts are not the cheapest of staff. If the instrumentation is there for warning purposes, the interpretation must be done frequently, and that is what pushes up the costs. Of course, alarms can be sounded automatically with a computer monitoring system, but those are not particularly cheap, and are subject to frequent false alarms caused by rogue readings. No doubt in time, artificial intelligence will reach the point when it can do the job as well as a human expert, but even when it can, it may be that a considerable time must elapse before it becomes cheaper as well.

Hutchinson had some piezometer instruments (see the following section) installed at Hadleigh Castle in Essex and maintained that it took 7 years for them to show a maximum reading. That was the maximum recorded reading over the 7 years, of course, and omitted the piezometric levels that occurred at any point between the periodic manual readings, and what happened subsequently. Of course, if the water levels in a piezometer reach ground level, and there is no reason to suspect that they can be artesian, then that maximum may well be true. The obviousness of this is obscured by read-outs that give the result in pressure units, and while in principle, pore water pressures are pressures, they are much easier to understand if they are given as pressure heads. Suctions, arguably, work better in pressure units for the simple reason that negative heads are difficult to visualise.

PIEZOMETERS

Piezometers are instruments for detecting the water pressures at a particular location in a soil or rock mass. The simplest type of piezometer is a plastic tube open at the top and with a porous tip at the bottom. The tube with tip attached is threaded into a borehole, sand is dropped or washed down the hole to form a pocket around the tip, and the rest of the hole is sealed, preferably with a cement-and-bentonite grout, but possibly with bentonite pellets. I don't advise the use of pellets unless the hole is 150mm diameter or preferably larger, and in wide boreholes some pellets can be used above the sand pocket to prevent it being disturbed by the grout. Boreholes that are water-filled and also wide do benefit from the use of pellets, because otherwise, grouting has to be done through a tremie pipe. Small clay pellets from augering in stiff clays, or pea gravel, can be used with effect to thicken a grout, but larger arisings usually get stuck partway down and make the backfill less effective, which is a euphemism for completely useless. The water level can be detected with the aid of a sensor, usually an electrical device that is battery powered and constructed so that when the water level is encountered, a circuit is made and a buzzer sounds. Measurement of water level is assisted if the cable is marked with depths. A tube is used instead of an open borehole so that water level changes require less flow into or out of the device, and so it responds more rapidly to groundwater level changes. Open wells can be used but are even slower to respond than open boreholes. Standpipe piezometers are jacketed by permeable material for their whole length. Automatic reading of such an instrument would be, for example, *via* a transducer installed in the tip. A simple piezometer like this has the advantage that if air bubbles enter the tip they rise up through the water column, and thus the device is self-de-airing. A typical inside diameter for the tube is 19mm (3/4 inch).

I have never seen good results from multiple piezometers in one borehole, and would never countenance their use in a project of mine: One piezometer, one borehole, is my rule. This applies even to expensive multibay instruments that, in my experience, are rather useless and a waste of the money to purchase, install, or read.

When installing the tubes in an open borehole, they are assembled from typically 3m lengths, with threaded or glued connectors. I have found threaded connectors to be best, especially when the length to be installed is large, as a simple glued joint is sometimes not strong enough and may separate during installation, thus needing to be taped as well for strength. The sand to form an enlarged pocket surrounding the porous tip is usually simply poured in from the surface and washed down the borehole with a little water. I had always regarded this procedure as likely to underperform, but on excavation at the Selborne test site, I was pleasantly surprised to see that nice sand pockets had been formed around piezometer tips by this crude method.

An electrical piezometer operates in much the same way as a transducer installed down a well or pipe, but with a porous tip and cavity in which the pressure on a diaphragm is measured. The volume changes to actuate the sensor are typically very tiny, and so the instrument responds very quickly. The pressures are commonly measured by means of sensing the vibration frequency of a tensioned steel wire attached to the diaphragm, and such piezometers are often called 'vibrating wire' instruments. Their biggest disadvantage is that they are not self de-airing, and so they are usually fitted with high air entry filters, the 'high' indicating the pressures required to force air in, not the amount of air that can enter. The filters and cavity need to be saturated before installing the device, with an electrical lead reaching up to the surface, where the instrument can be read from a gauge house or portable readout, either of which can connect to a datalogger and be read remotely through a cell phone or satellite link.

Pneumatic piezometers have the same type of high air entry tip and cavity, with a diaphragm of much greater flexibility than in the electrical piezometer. The diaphragm covers two openings for twin air lines. Air (or some other gas, commonly nitrogen) is supplied down one line until the diaphragm lifts at a pressure corresponding to a little more than the water pressure in the piezometer cavity, and this allows the air to enter the second line where the opening pressure is registered. Then, when the pressure in the first line is reduced, the diaphragm closes off that second line, giving a closure pressure. The mean of the two readings is the piezometric pressure. Again, the leads come up the borehole through the grout, and can be used with a portable readout or be permanently connected to a reading system.

A fourth type of piezometer is the hydraulic piezometer, which, like the pneumatic device, has two tubes, one of which connects to the top of the cavity in the piezometer's installed upright position, while the other continues to the bottom of the cavity. This arrangement ensures that air in the cavity can be flushed out. Indeed, the whole system can be flushed to remove air, including both tubes. Since the water-filled tubes are exposed at ground level they need to be provided with an antifreeze solution in de-aired water. The pressure at the piezometer tip can be read by measuring the pressure at the end of one of the leads and making allowance for the elevation difference between the piezometer tip and the readout, and this limits the allowable separation between the tip and the readout to less than 10m, preferably much less. This limitation makes the system better at being used in embankment fills, as the read-outs can be on the embankment face at the same elevation as the tip, with the leads laid in a carefully backfilled trench. Tips installed for this purpose are usually tapered and installed, not in a sand pocket, but directly in contact with the soil in a specially formed hole into which the tapered shape of the tip fits exactly. Traditionally they were read with the aid of a mercury manometer, but any pressure readout ranging from a Bourdon gauge to an electrical pressure transducer can be used, with the latter better for remote reading or datalogging.

Piezometers in a landslide are best if positioned *on* the slip surface, and because I don't like an initially grouted section with a piezometer tip above it, it does mean that I prefer a second borehole alongside the one in which the slip surface position has been identified. A good approach is to drill down to within a metre of the anticipated depth, and then to sample the soil with immediate on-site logging, so as to position the piezometer in the second hole precisely. The original, fully sampled or cored borehole is available for an instrument that goes deeper than the slip surface, which might be a deep piezometer or perhaps an inclinometer. The piezometric pressure recorded at the slip surface of a landslide is directly applicable to stability analysis, but a further deeper or shallower piezometer indicates the direction of groundwater flows, with the shallower one(s) providing a potential insight into how water infiltrates from the surface. Unlike in the detection of the slip surface *slope*, where certain arrangements of the boreholes are more helpful than others, the arrangement of boreholes for piezometers at any particular location is less critical. In the Selborne test site they had to be arranged in a line along a narrow bench in the slope: the boreholes were drilled, and the instruments installed before the next lower slope facet was cut.

In a slope that has not failed but for which the distribution of piezometers with depth is such that a profile of piezometric conditions can be established, the sequencing of boreholes with different hole depths starting with the deepest may not be the best approach, particularly if the deepest hole is to contain an inclinometer casing. Installing the shallower piezometers first may make it easier to capture the early stages of piezometer response while a crew is at that location doing the deeper hole with all the sampling. However, doing the deep sampled hole first may help establish the particular horizons in which placement of piezometer tips is advantageous.

It is, in my opinion, essential to capture the initial equilibration phase in a piezometer, as this gives some indication of the permeability and/or compressibility of the soil, and allows the final equilibrium piezometric level to be estimated without necessarily waiting for the piezometer to indicate it. Thereafter, the piezometer can be used for permeability testing, although probably not where it has a high air entry filter tip, because then it is probable that the low permeability of the tip influences the result. Twin-tube hydraulic piezometers with low air entry tips are ideal for the constant head permeability test,[10] and tube piezometers are good for falling and rising head tests.

A plot of water pressures *v.* depth for several piezometers in separate boreholes at the same location gives you an idea of whether there are elements of upwards or downwards flow.

A phenomenon that besets some piezometers is that there is a rapid rise between some readings followed by a slow fall. With tube piezometers this behaviour is the result of water ingress, and it is a common problem with instruments finished at ground level in a stopcock cover. Basically, it is water entering the tube from the surface. Piezometer tubes should, in my experience, finish above ground level for this not to happen.

The problem with two piezometers in the same borehole with an inadequate seal is revealed when the two show exactly the same water level and rise and fall in step. It can also be the case if the upper piezometer has no reading until the lower one catches up, and this shows a completely inadequate seal.

INCLINOMETERS

The inclinometer system is a combination of a tube installed in a borehole and a readout device consisting of a probe that can measure the inclination of the tube to the vertical. The simplest design of probe has sprung wheels to allow it to be lowered down the tube, with the wheels running in keyways formed in the inside of the tube. Tubes have four keyways, and early probes could only measure in one plane. There were two ways to insert the instrument to read in that plane, and the two sets of readings could be averaged or compared to check for gross errors. Then the probe was used to measure the shape in the other plane. Modern instruments are biaxial, which means that they read in two orthogonal vertical planes simultaneously, and this simplifies matters, but it does mean that operators skip the check positions, leading sometimes to odd errors. However, modern instruments are much more reliable than early-generation devices. Read-outs in particular can now datalog the readings and accumulate them.

A development is to have a string of sensors installed in the access tube permanently, with datalogging and remote readout so that the readings can be taken in real time. While these instruments are hugely expensive, they can be pulled out of the casing at the end of a project and be reused elsewhere.

The classic method of presenting the data is to draw the shapes of the inclinometer tubes at the various reading dates, perhaps in different colours, and then to compare those shapes. Ideally, a step should occur where the inclinometer tube crosses a slip surface. This approach makes it difficult to see what is happening if there are too many sets of readings. A useful alternative is to show the time history of deflection for each inclinometer segment. This is best viewed on a computer monitor, as otherwise, one gets overwhelmed with paper. Real time reading of inclinometers does provide the person whose job it is to interpret them with an embarrassment of riches, and getting the computer to plot the data does help you avoid becoming overwhelmed. Large datasets also assist some of the analysis of the data; for example, it is a simple matter to derive the statistics for the readings, so that alerts are only given for changes outside the expected range. If a particular inclinometer segment is showing increasing deflections with time which suggest that a failure is developing, a plot of the inverse of the rate of displacement (y axis) relative to time (x axis) in many instances will follow a more or less straight line that descends and eventually crosses the time axis, thus predicting the occurrence of a runaway failure. The technique is called a Fukuzono or Saito plot.[11] It works best and most reliably when the factors leading to failure are constant: if they stop and this causes the deformations to stop increasing, then the predicted axis-crossing time of failure can no longer apply.

LABORATORY TESTING

Much laboratory testing is a waste of time and money. The problem lies not so much in the laboratory (although sometimes one wonders what they actually did) but in the way that samples are taken, preserved (or not), transported and handled. More to the point, you usually don't realise what was useful and what was not until long after the dust has settled, and the paperwork has been completed. You rarely realise at the time what was going to prove of little value, and laboratory testing is generally specified in good faith with some expectation in what it will demonstrate. If you must become an adept at laboratory testing your training will include reading the three-volume set of books by Head & Epps.[12] Those books are infinitely more readable than the appropriate Standards.

A fundamental problem is often at the outset: for example, when preparing a contract, you simply do not know what tests will be required, or if samples good enough to test will be obtained in critical locations. A laboratory testing schedule normally indicates the preferred prices that a contractor will charge, and sometimes, that is what you get in a tender, but equally, I have known cases where a contractor has loaded the items which you have asked for price only, but where he knows that those tests *will* be required and thus increasing his likely profits without increasing his tender price. Equally, he may under-price things he thinks will not be required, as this turns the tender price to his advantage.

In my experience, all landslides in clay strata will need some sort of residual strength testing, and as disturbed samples are always available, a test on remoulded materials such as the ring shear test presents itself as an obvious choice. The ring shear test[13] takes comparatively little soil, as the sample container is small and the test can be multistaged without the effects of progressive failure that multistaging has on the determination of peak strengths, as at residual strength the soil has no remnant brittleness. I would only call for a drained test on a pre-existing slip surface if I knew that the ground investigation had a means of collecting block samples, as often, by the time that you have found a slip surface in a core, the core is no longer suitable for testing. A 'price only' is always useful, however.

In a limit equilibrium method applied to unsheared soils, measurement of the residual strength is unnecessary, except to frighten the analyst into an awareness of what might happen if a failure does occur, but in a continuum method, it is the ultimate result of local overstress and is probably needed in an analysis of progressive failure. Since most slopes have some overstress somewhere, you probably need to understand the peak-to-residual stress-strain behaviour as well.

In specifying the stress range for tests on slip surfaces, whether sampled in the field or formed in the laboratory, a range of stresses needs to be specified. I often find that very low normal effective stresses give an imprecise result, and therefore favour trying to avoid them. I prefer to specify a normal effective stress for a single stage that is more or less the vertical effective stress at the sample location, and for a multistage ring shear test, a little less than that in-situ normal effective stress up to about twice or three times as much, taking into account what lowered ground water pressures there might be due to drainage in the stabilisation design, or increased normal stresses due to loads from fills or anchors. Usually, these figures are guesses at the time tests are ordered.

The residual strength is most useful for only one purpose, and that is for stability analyses if the whole or a large part of a slip surface is left in-situ after remedial works have been carried out. It won't be of much use, say, if the whole landslide is dug out. I prefer using laboratory-measured strengths to confirm and extend what I get from back analysis (see Chapter 4) rather than to base my design on the laboratory-measured properties alone. Measuring the index properties and perhaps also moisture content helps in understanding the differences between successive tests.

The ring shear test cannot usually accommodate anything with a particle size larger than sand, and gritty soils wear away the necessary roughness on the platens and should be avoided if possible. For those gritty soils I would prefer to use a shearbox with a pre-cut plane pseudo slip surface formed in the laboratory. Ring shear tests are also not much good for determining the peak strength of anything, because at small rotations the strains are larger at the outside than the inside. It works for residual strength because at large rotations the strains have reduced the soil to residual strength right across the specimen.

For peak strengths we usually have a choice between a shearbox (direct shear) or a triaxial test. The advantage of the former is mechanical simplicity, and of the latter, the ease with which drainage can be controlled and the ability to follow different stress paths. The strains in a triaxial test are also not so constrained as in the shearbox, and this is often felt to lead to a better result.

Effective stress triaxial testing is both time-consuming and expensive, with consolidated undrained testing with pore pressure measurement slightly quicker but more technically demanding than consolidated drained testing. I usually find myself unsatisfied by the saturation and consolidation stages in these tests, because a saturation stage that does not involve application of back pressure can result in excessive effective stresses and a laboratory overconsolidation effect, whereas the application of back pressures usually causes an increase in water content that affects the moduli controlling deformation and may also mean that the consolidation stage is largely a removal of water introduced previously.

As for *undrained* peak strengths, the 'quick' unconsolidated undrained triaxial test is cheap and has its uses. Undrained tests can also be carried out in the shearbox, but as the only real control on drainage is the rate of shearing, and the pore water pressures in the sample cannot be measured in the standard variation of the test, it is not really suited to undrained tests on anything except clays, although consolidated undrained tests are possible. A little-understood problem with a consolidated undrained triaxial test in a dilatant soil is that the pore water pressures may decrease during the shearing stage to below the value required for full saturation, leading to the formation of air bubbles.

In practice, peak strength testing is fraught with problems. Firstly, even if money is no object, effective stress testing is slow, and results may not become available until long after samples were first taken. Even laboratories that specialise in effective stress triaxial testing only have a finite number of cells and test frames, and for reasons of economics, there is always a backlog of tests for each piece, further adding to the slowness of the process. There is also the problem of representativeness, *i.e.* of testing samples from critical places. This isn't a problem with residual strength testing, as tests are usually done on soils either side of a recognisable slip surface and not randomly through the soil mass. If they *are* done randomly through the soil mass, then do not expect them to have much bearing on the problem at hand.

The point is that where slip surfaces develop might be the result of peak strength rather than residual strength, and soil that is weak after shearing is not necessarily weak before shearing. Also, there will be influences such as relative shear moduli – which may or may not reflect peak or residual strength, whatever else there is in the slope that affects strains, how the slope is formed and, of course, what are the pore water pressures, and those may be far from constant as the failure develops or propagates.

The precise results you get from testing are also strongly influenced by not only how the samples are taken, but also how they are handled, protected, and transported. Typical faults in these activities include allowing the samples to dry out or be subject to frost (including during air transport), how cores are extruded from core barrels, whether cores are cut up or broken to fit in short core boxes, whether they are bumped around while taking them from the drill rig to the laboratory *via* site offices, and how long they sit in store, waiting to be examined or tested. The biggest effects of all of these problems are on the c' element of strength, with ϕ' relatively unaffected and much the same on a remoulded or reconstituted sample, so in general I prefer not to trust laboratory c' values and also rather like testing remoulded samples.

Natural deposits also contain structures such as sedimentary sequences, and thus the results will span a range of strengths anyway.

NOTES

1 See Hutchinson, J. N. (1983) Methods of locating slip surfaces in landslides. *Bulletin of the Association of Engineering Geologists*, *20* (3), 235–252, 1988.
2 To which pronouncement, some other geotechnical professional will declare it to be utter bunkum, as there is no such difference, whereas another will say that the classification is far too coarse and that there are hundreds of different

types of geotechnical engineer, and not only that, but the proper classification was produced by some obscure author in the 1930s. All in all, a bit like landslide classification, really.

3 A leading geotechnical engineer in Italy, of all places, once remarked to me, "*Eddie, you know a lot of Geologists*". I did not know whether it was a compliment or an admonishment. I still don't.

4 You may also find that various other tools excite the attention of Security at airports, including the laser spot on a presentation aid, a bladed multitool, especially a Swiss Army knife or, as I once discovered, a retractable tape measure. I have cogitated at length (no pun intended) about whether or not I would be intimidated by someone bearing a tape measure, and eventually, I have concluded that sometimes in the course of daily life, one meets complete idiots.

5 I suggest that you also learn from colleagues, as their experience will certainly be different to yours, even if they are your juniors.

6 I have used a shell-and-auger rig and think that it is positively dangerous. I have also used a top-drive hydraulic rotary rig for drilling and instrumentation and used it for rotary coring down to 85m, although it could only auger with continuous flight augers to 45m. By *used*, I mean that I have driven the rig. It is actually hard work.

7 Nowhere better than in Norbury, D. R. (2015) *Soil and Rock Description in Engineering Practice*. Second Edition. Whittles Publishing.

8 The case I am thinking about is particularly egregious, since I logged the cores! The fanciful ground model even made it into a publication in a respected journal.

9 See Dunnicliff, J. (2008) *Geotechnical Instrumentation for Monitoring Field Performance*. Third Edition. Wiley.

10 See Gibson, R. E. (1966) A note on the constant head test to measure soil permeability in situ. *Géotechnique*, *16* (3), 256–259; Gibson, R. E. (1970) An extension to the theory of the constant head in situ permeability test. *Géotechnique*, *20* (2), 193–197, and also Bromhead, E. N. (1996) Interpretation of constant head in situ permeability tests in soil zones of finite extent. *Géotechnique*, *46* (1), 133–143.

11 Fukuzono, T. (1985) A method to predict the time of slope failure caused by rainfall using the inverse number of velocity of surface displacement. *Journal of Japan Landslides Society*, *23* (2), 8–13; Saito, M. (1965) Forecasting the time of occurrence of a slope failure. *Proceedings of the Sixth International Conference on Soil Mechanics and Foundation Engineering Montreal. University of Toronto Press. Toronto*, 2, 537–541. There are many other papers on this topic.

12 Head, K. H. & Epps, R. J. (various dates) *Manual of Soil Laboratory Testing* (3 vols). Whittles Publishing. Head wrote the original and produced a revised second edition, with Epps updating things to produce the third edition. When I entered the profession, the best book on the subject was by T. N. W. Ackroyd, all in Imperial units and now a museum piece. Head & Epps's books are a million times better.

13 BS1377 lists a ring shear test in the small apparatus that I invented as a total stress test. I suppose that it is, but it is an effective stress test with zero pore water pressure, and it yields parameters in terms of effective stress. Sometimes one wonders what one's contemporaries are drinking, smoking or injecting, or whether they are simply idiots.

Chapter 3

Things that cause landslides to occur or move, from geology to human stupidity

THE GEOLOGICAL STRUCTURE

If it had not been for plate tectonics, we would probably not be the slightest bit bothered by landslides, because the sea covers about two-thirds of the planet, and its average depth is greater than the average height above sea level of the continental land masses. There might be the odd volcanic sea mount, but that's about all. Humanoids, if such creatures existed, would be mer-people. Plate tectonics shoves mountains skywards and creates the gradients in rivers that transport landslide-sourced sediments down to the sea. The process acting over geological timescales creates continents and sedimentary basins in which those sediments can accumulate to make new rocks, so that we aren't limited to the igneous rocks that geologists have given a myriad of names to for the confusion of civil engineers (who, after all, did a lot to invent geology in the first place).

We probably wouldn't have ice caps, as there wouldn't be a land-locked polar sea or the other pole on a landmass, and we probably wouldn't have glaciation either. It would be *Waterworld*.[1]

But we have plate tectonics, and we do have continental landmasses and islands, and we do have vulcanism, and we do have igneous, sedimentary and metamorphic rocks. We do have erosion, and we do have uplift, and in combination, we have the setting in which, over most of the landmass, we get landslides.

There are, however, certain geological settings in which we find landslides. The first and most obvious place to me is on the coast, where the sea erodes and undercuts, and when the coastal slope is steep enough, you get landslides. What sort of landslides you get depends on what rocks are present, and the frequency and activity of the landslides depends on how strong the rocks are, how intense and continuous marine attack is, and also, how wet is the weather. Strong rocks fail with falls. Strong rocks with an erodible outcrop at around sea level give you topples. Seawards dipping sedimentary strata with weaker beds in the sequence give you slides. Indeed, even if the strata are horizontal, or even dip gently inland, you can get slides. They tend to be compound in section, with slip surfaces developed along the weak

DOI: 10.1201/9781003428169-3

layers. It was landslides like this that I cut my teeth on, because in the south of England, where I live, the geology is a mixture of sedimentary rocks of Mesozoic and Tertiary age, with mainly gentle folding that typically gives low-angle dips.

Southern England is hardly mountainous, and even the North, Scotland and Wales have barely any peaks that would count as mountains anywhere else in the world, although the natives of the UK count the big hills as mountains. But the mountain ranges of the world, many of them still being pushed up today, are busy eroding as I write, and given the odd strong-motion earthquake combined with erosion by rivers and glaciers and with rocks that are sheared and fractured, with sedimentary sequences tilted to all manner of shapes and inclinations, you have another very obvious setting for landslides. Not that most of them mattered much to humanity except in the last century or so, because the population those mountains supported was rather small, but nowadays they are sites for dams and recreation and so are more densely populated than would have seemed possible in antiquity.

Taking just those two environments, you will find some similarities but a huge range of differences. Some of those differences come from how the various slopes evolved, and this, in many cases, includes accounting for the effects of various climatic processes.

Critical factors occurring due to tectonics are that strata can become folded or faulted. Faults are little different to pre-existing slip surfaces, as they represent a weakness in the rock mass that can be exploited by a subsequent landslide. Folds may involve shear along beds, or if they have an axis which is downwards and outwards relative to a topographic slope (say a coastal cliff), they increase the susceptibility to landsliding. However, in the right circumstances, failure can even propagate *up* slight dips.

Landsliding inevitably picks out weak strata in any geological succession. Weak strata may simply result from different environmental conditions and the supply of sediment at the time of deposition, in which case there may be alternations of mudstones with limestones or sandstones, but occasional injections of clays from volcanic ash create thin slide-prone horizons that are difficult to detect with the human eye, but that landslides seem to have no difficulty seeking out! Volcanic ashes appear to be readily converted to smectites assisted by the gut processes of bottom-dwelling organisms.

GEOLOGICAL HISTORY

In the north and midlands of Britain, geological history almost certainly includes the results of glaciation, and on the coast anywhere, it may mean past erosion by the sea, but also the accretion of coastal deposits in spits, bars, beaches and, in some locations, saltmarshes and estuarine sediments. In rivers, the variations of past sea levels cause responses to the line of the thalweg as the base level of the river changes and the cutting down of the river through previous floodplains leaving terraces. Landslides in the low uplands of the UK commonly involve glacial deposits.

Rivers are sometimes blocked by landslides, and then a lake forms. If the resulting landslide dam is overtopped or undermined by seepage erosion, it may give way and release a flood, but it also leaves characteristic landforms: the lake-bottom sediments as a terrace, some meandering rapids where the rocky core of the landslide dam has been cut through and, where the channel is forced to develop at one side of the original valley, stimulated instability on that side.

Sediments that are buried under great depths by later deposits and then uncovered by erosion show strength gains due to overconsolidation. The stress relief commonly manifests itself by the development of fissures in the soil, giving rise to stiff, fissured, clays.

Superimposed on this, there is the erosional pattern of the Quaternary, with deposits from that era. Relative to the land area, the UK has a long coastline, with active erosion, and good outcrops, including many areas of coastal landsliding. Landslides also occur inland. There is no present-day vulcanism in the UK, and earthquakes are mainly of an undetectably low magnitude. Clearly, this 'specification' does not apply elsewhere in the world, and there are places that were never glaciated but where tropical weathering and erosion have taken place for perhaps millions of years.

As plate tectonics shuffles the landmasses around, deserts appear in the geological record that cannot occur there today, and perhaps glacial deposits where there are now deserts. As well as plate tectonics, we also have climate and sea-level change, most instances of which have absolutely nothing to do with humankind or fossil fuels.

THE WEATHER

Rain is a big factor in landslide *activity* but not necessarily in landslide *occurrence*, which is largely influenced by the geology and erosion. The distribution of landslides in the UK mirrors the outcrops of clay soils and also follows the coast, with a subsidiary set of landslides inland where human activity – infrastructure routes, mining, and quarrying – has caused or stimulated instability. Given the pattern of rainfall, which is highest where the rocks are strongest, perhaps this distribution in the UK is unsurprising, but it may not be mirrored elsewhere.

Snowmelt has a tendency to be different to rainfall in terms of infiltration or runoff, with hail having virtually no effect at all. The spatial patterns of rainfall as storms move across country may not be benign if river flows are consequently turned into erosive floods. Intense rainfall in tropical storms, typhoons and the onset of monsoon rainfall also has dual effects of infiltration and erosion. There is a tendency amongst the citizens of the UK to think of the rainy season as being in the winter,[2] but in some locations, it is during the summer or *monsoon season*.

In hotter countries, the diurnal temperature changes with expansion and contraction of rock faces may lead to the detachment of blocks from rock faces, but where there is freezing, the expansion of water in joints can also prise off blocks from a face. Even if ice only forms at the surface, it may lead to a rise in pressures in water-filled joint networks, as the water cannot escape from ice-blocked outlets.

RIVERS AND THE COAST

Rivers have the habit of lowering their thalweg,[3] and this process undercuts and destabilises riverbanks. Rivers and streams also meander where the gradient of the thalweg is gentle, alternately eroding the valley sides and thus causing fresh landslides, or moving away from the valley sides so that the slopes are 'abandoned'. After such abandonment, the slopes may continue to fail or move, subsequently reaching a state of relative stability – until exceptional weather or human activity provokes reactivation along the slip surfaces that remain inside the slope.

Rivers respond with accelerated downcutting if sea level drops in a glacial episode but respond with sedimentation when sea level rises again and drowns out the estuary. This may leave landslides perched in the hillslopes or interdigitated with sediments.

Changes in relative sea level and the growth of coastal marshes or spits may cause coastal slopes to become abandoned in the same way that inland slopes are abandoned by river meanders, with the same sorts of hidden potential to be easily destabilised.

It was at a conference in September 1978 that Professor Hutchinson mused, tongue in cheek, that as the quality of British architecture was so abysmal near the coast, was it in fact right to see coastal erosion as an entirely negative natural process? Unfortunately, if one is not an architectural critic but a landslide engineer, the answer to the question is that coastal erosion is always a negative factor. The sea scours away at the toe of coastal slopes and causes erosion at a variety of intensities, depending in part on how energetic the processes of wave and current action are and how susceptible to erosion the coastal slope is. Wave activity is a function, in part, of weather and how exposed or sheltered the coast is, for example, in a bay approached by a wave train obliquely, the waves may be refracted by a headland so as to concentrate attack in one place, and minimise it in another. The long-term effects of this variability are usually apparent in the plan form of the coastline, which integrates the effects of long-term, multiple-event, erosion, and its interplay with geological structure.

In many coastal locations, the wind direction commonly has both a primary and a secondary direction, with the frequent storms from one direction with the longest fetch causing regular erosion that we become accustomed to, and build coastal defences to prevent, only to be surprised when the inevitable storm coming from a different direction strips away the beach and causes erosion in a day or so that should have been anticipated, but was not. Hutton's principle of uniformitarianism does not imply that every day is the same!

There is a certain prejudice about the 'end groyne' problem in that coastal erosion is thought to be accentuated after the last groyne in the predominant drift direction. While it is true that the purpose of groynes is to reduce longshore drift, and that will undoubtedly cause beach denudation downdrift (if they work as intended), every case I have inspected shows that the coast defences have worked, but beyond the 'end groyne', erosion and coastline retreat goes on as it did even in the defended section prior to the works, and the resulting difference is coastline alignment is the result of the coast defences working, not in their having a deleterious effect on the neighbouring stretch of coastline downdrift. Coastal defences do, however, have the effect of reducing the sediment supply for the coast, and some other works are needed to retain beaches.

Just as in the case of river erosion, coastal slopes may become abandoned, usually due to the development of saltmarshes, or in the tropics, the growth of mangrove swamps.

EARTHQUAKES AND VOLCANOES

Earthquakes trigger landslides in seismically susceptible areas, and apparently, more so if the earthquake happens in a rainy season. Strong-motion earthquakes happen most at plate boundaries and therefore in or close to mountain belts. At the present level of technology, there's nothing that can be done about them. Lava flows are not readily destabilised when they have cooled, but deposits of ash are.

A surprising number of extrusive volcanic processes are, in effect, landslide processes, from volcanic bombs that might have a different source mechanism, but land in much the same way as a rockfall with the only difference being that the volcanic bomb might be very hot or even molten! Like a rockfall, there is very little an engineer can do except keep clear until it stops.

You could also say that *nuées ardentes* or pyroclastic flows are basically hot flows of silt-sized particles or that streams of lava are hot debris flows.

However, the most obvious landslide mechanisms of volcanoes are when the edifice fails due to internal pressurisation or through being 'built' on a weak foundation.[4]

An interesting feature of natural landslides of the landscape feature kind is that they often have a Factor of Safety marginally above 1. The additional stability is because they last moved when the groundwater table was higher or, in landslides in a seismic setting, when the last strong-motion earthquake happened, especially if that was at a time of adverse groundwater conditions. A net result of that historical coincidence is that it is usually possible to make a start on works without provoking an immediate disastrous movement, but instead, if the signs are ignored, it occurs when a workforce is committed. I always advise caution where work on identified landslides is concerned.

MY FAVOURITE LANDSLIDE TYPES AND SETTINGS

Inevitably, if you have had to deal with a lot of landslides, you find favourites. I have no doubt that the late Prof Hutchinson was intrigued by the problems of abandoned and defended slopes, because he encountered so many in his survey of coastal landslides. He spent a lot of time working on mudslides, and he found the dating of landslides of particular interest. As for myself, I generally prefer – if I have any choice – to work on compound landslides with a major part of the slip surface running along a slide-prone horizon. The setting of gently dipping sedimentary sequences with stiff clays is one of my ideals. It's the 'home job'.

Then there are geological settings that I find particularly interesting. One of them is a sequence of tills. In general, I dislike writing the word 'till' in my notebook, as a site scribble may be read back as 'fill'.[5] Sometimes I have encountered a sequence of two or more tills overlaid on top of each other and with a laminated deposit in between. Where this deposit has clay laminae, they can act as slide-prone horizons, but if they are laminated silts and sands, they do not but instead become susceptible to seepage erosion (Chapter 7). Two different tills may be related to different phases of glaciation with materials originating from different sources, or the lower one might be a *lodgement till* emplaced at the base of a glacier and the upper one a *meltout* or *ablation till* left as the glacier thawed. Despite attending several conferences on the engineering geology and geotechnics of glacial materials, I never did discover the mechanism that emplaces those laminated deposits, although elsewhere, they are associated with seasonal variations in the sediment supply to lakes. I was, however, introduced to two words: *allochthonous* and *autochthonous*, words that give the same impression on the listener as *supercalifragilisticexpialidocious*![6]

They simply mean 'originated elsewhere' and 'originated where it is now' respectively, and it is the former that I have found most intriguing. Tills almost by definition are allochthonous, but it is when the solid geology is *allochthonous* that we get a setting in which landslides may develop.

Allochthonous rocks are generally those that have been thrust by tectonics many kilometres from their place of origin by low-angle thrust faults. Most textbooks on structural geology illustrate this with a diagram including a huge displacement thrust fault and comment that at the 'snout' the rock mass may be broken up into isolated hills named 'klippe' (from the German, plural *Klippen*) but fail to point out the engineering significance, and that is not when the original stratigraphy is preserved, but that the rocks may easily be overturned as they are in nappes, nor that if the strata above or below the thrust fault may be sheared in a chaotic fashion. Clay soils then tend to develop a 'scaly clay' fabric (*argile scaglioso* in Italian) where the soil is not simply fissured like we find it in the UK but sheared into lens-shaped pieces with slickensided surfaces top and bottom. Limestone and sandstone strata within the clays are whipped up into olistoliths that pepper the landscape looking like archaeological sites or ruined buildings.

If the strata are silty rather than clayey the scaly clay fabric isn't developed, but the chaotic multidirectional shearing leads to systems of faults along which instability may develop and the inclusion of just about anything that is in the way. These inclusions also get sheared in the process, and also have sheared surfaces all around them, so that cutting through one of them can easily release a landslide. It is the use of the word *allochthonous* by a geologist in this context that catches my attention.

ANIMALS

Sometimes burrowing animals get surprises, as in the case of a badger sett dug into the bank of a canal. The animals got sufficiently far in for seepage to do the rest, and the water in a length of canal issued through the badgers' burrows, eroding them to a larger size at first, then collapsing the tunnel roofs and eventually carving a slot into the canal bank, causing flooding in the surrounding land. Was it a landslide? Probably, at one stage or other.

Normally animals (and plants) do not cause landslides, but they can inhibit stabilisation works, typically by being a rare or protected species that may not be disturbed. Plants such as the Japanese Knotweed prevent you from disposing of excavated soil without a lot of expensive treatment.

The other burrowing animal that I have found fascinating is the mole. On a site I inspected there was what seemed to be a network of cracks in the ground, possibly due to shrinkage but, more worrying, that they were possibly cracks due to ground movement. It turned out that they were, in fact, shallow mole burrows that had collapsed, tracing out on the ground the irregular hunting pattern of moles looking for a meal!

Probably the animals that most often cause or reactivate landslides are of the human kind! What they do and how they do it is covered in the following sections.

THE ANTI-BERM – DIGGING AWAY AT THE TOE

One sees this at various levels of stupidity, and it almost always leads to problems. There are cases where the excavator did not know he was digging at the toe of a landslide and set off movements of various degrees of severity, but there have been cases where despite being warned in advance, an obstinate or foolish contractor has gone ahead regardless of the consequences.

You sometimes see the instruction about excavating a trench for it to be done in short sections. This is no doubt good advice if the short sections are sufficiently short as to only open a small percentage of the width of a landslide at a time, so for instance, I should rarely worry about a trial pit at the toe of a landslide, but there are practical limits. For example, the case where an excavation into a river bank large enough to contain a pumping station[7] released a landslide that had a footprint much larger than that of the excavation, and which was not detected because the slopes contained no sign of previous landsliding on the site.

Elsewhere (Chapter 9) I have made a note about benching in before placing a toe berm, and this has been recorded as precipitating landslide movements.

Incidentally, when preparing a site to receive fill, stiff clay successions benefit from a topsoil strip and removal of the weak, weathered, upper layers, but recent deposits with a stiffer, weathered crust do not.

THE ANTI-DRAIN – RECHARGING THE GROUND WATER

The most obvious groundwater recharge mechanism is to put water into the ground deliberately, which is what is done with soakaways and septic tanks. Soakaways typically take runoff from roofs or paved areas and put it in the ground to 'soak away'. Rainwater put into a soakaway does not have the prospect of evaporating under the combined effect of sunshine and wind, or on a slope of running off when the precipitation exceeds the capacity of the ground to absorb it, but once in a soakaway, the water continues to infiltrate long after the rain has stopped. Soakaways located near the crest of a slope are most obviously damaging.

The same goes for septic tank sewerage, sometimes called a cess pit[8] in the UK. This is where the wastewater from a building is collected in a pit where solids settle out, and the supernatant (lovely word, that) water is channelled away in pipes to soak into the ground. Ideally, the human gut bacteria are defeated by the soil bacteria, but it is possible to overload nature's way of dealing with such things and contaminate the groundwater as well as vastly increasing the infiltration rate, and making the groundwater levels rise. If the contaminated water is drained away, it needs treatment to remove the pollutants prior to disposal. Site workers need to be made to use gloves and other measures to avoid contaminants.

Modern lavatories use sometimes as little as 5 litres per flush, which may be a half to a third of older installations and are therefore helpful to reduce the load on septic tanks. However, if they are inefficient and require multiple flushes, they may effect little improvement. Similarly, a shower consumes less water than a bath, but the modern habit of showering every day at least once may consume nearly as much water as our grandparents' weekly bath. Also, a modern, plumbed-in, washing machine may use rather more water than its users believe.

Shallow land drains or field drains collect infiltration and transport it, usually to ditches at field boundaries. In some cases, drains were made with tiles in an inverted Vee, or later with butt-jointed earthenware pipes. They certainly help make land less soggy and waterlogged, and improve a field for animal husbandry. Clearly, if used in a field that is later ploughed, they would be destroyed. The intermediate case, where they are blocked, is worse than having no drains at all, because water can back up and reach artesian pressure levels. Open ditches, of course, may overflow, but do not become artesian. Drains may become blocked through siltation, or being shallow, they can be disrupted by vehicle wheel ruts. Another form of blockage is when field drains are encountered during a ground investigation: trial pits are notorious disruptors of drains, especially if the drain run is not reinstated when the pit is backfilled. Fortunately, property developers building housing developments on former agricultural land generally substitute services laid on pea gravel for the disrupted land drains.

I have experience of both soakaways and blocked drains causing instability. In several cases the soakaway or septic tank discharged near the head of a slope, but I have also seen septic tanks discharging into pre-existing landslides and reactivating them.

Two interesting cases involved the deliberate blocking of drains by building contractors. In the first case, a farmer had a retirement bungalow built at the bottom of a field on his land, in the only position for which he could get planning permission. To create a level site, the building contractor had to cut in at the toe of the slope and, in doing so, severed a drain. At some point, the water issuing from the drain became inconvenient, and so the drain was deliberately blocked. Over the next few years, a slide developed along the alignment of the field drain, spilling debris against the rear wall of the house. Fortunately, the farmer had access to some plant and had cleared the debris several times before the obvious solution, to reinstate a drain up the axis of the shallow landslide, was installed. This stopped the slide in its tracks. Active landsliding was seen in the adjacent field, where perhaps field drains had never been installed or where they had been neglected.

In the other case, land drains in sloping ground had been buried by the earth fill for an estate road. It was later decided to build a school with sports pitches on adjacent land. The pitches were built with cut and fill, during the course of which numerous land drains were severed. Again, the discharges in wet weather were an inconvenience, and the contractor blocked many of the drains with clay. Then, with heavy rain, the drains, which were by now buried to a depth of some metres, fed water into the ground, which failed the sports pitch cut slope and the overlying highway embankment.

Water is also imported, unseen, along the bedding for service pipes and pipelines, and a combination of bentonite plugs and outlets may be needed to prevent it reaching vulnerable slopes. In urban areas, there are service trenches for gas and electricity as well as for water-bearing services such as the delivery of potable water and removal of wastewater. A third system of seawater delivery for sanitation is also sometimes used in locations without a good, reliable, treatable water supply, such as Hong Kong. Evidently, the saline water is used for lavatory flushing, and especially when mixed in the waste stream with non-saline water used for washing, the saline content does not interfere with wastewater treatment. However, having three instead of two distribution systems, two of them with service reservoirs and all liable to leakages, represents an additional potential source for adverse effects on the natural groundwater body.

THE ANTI-EXCAVATION – LOADING AT THE HEAD

In some cases, a householder with a landslide will have a perspective from his property that is different from an engineer working with a surveyed cross section and a mental picture of a slip surface and movements along that surface involving rotation. Instead, the householder sees a hole to be filled that will reinstate his garden or whatever. This was the case where a coastal landslide[9] had carried away a combination of houses and gardens, followed over more than a decade by gradual collapses of the rear scarp until the garden of a house that had been unaffected by the original landslide was then teetering on the edge. In that case, filling the 'hole' was arranged, and a demolition contractor obligingly supplied lorry load after lorry load of bricks, concrete, pipes, broken bathroom furnishings, tiles and reinforcing bar offcuts until the local authority got wind of the illegal tipping and served a prohibition order. By then, the localised loading had caused the much more extensive underlying landslide to move, increasing the height of the rear scarp and provoking a new round of collapses of that heightened rear scarp. It was an unprotected slope, and no remediation could be done for environmental reasons, and in the event, that householder hastened the demise of his own property.

In another case, the loading was not at the head of the slope, but somewhere in the middle. This particular slope[10] was mantled by shallow landslides and was to be crossed more or less along the contours by a dual carriageway road mainly supported on an earthfill, which naturally was designed to reach its maximum height on the downslope side. At about one-third of the design height, the fill caused movements to occur along the old slip surfaces. It was not the whole of the fill that moved, but only the part of the fill with the main slope batter. In this case, stability was finally obtained by separating the carriageways so as to minimise the quantity of earthworks required, combined with drainage introduced into the pre-existing landslides. However, this happy state was not achieved without a further gaffe, which was to excavate away the fill, and dig a huge trench into the hillside along the line of the road – which, of course, released the landslides uphill of the road alignment. The final completed road with its separated carriageways was constructed on a new alignment.

Loading at the head of a slope also provides the possibility for the fill to fail without the underlying slope, and instead, the debris simply slides or flows over the pre-existing topographic surface. Such was the case in the Cilfynnydd coal mine waste tip failure in 1939 that blocked a road and a river, and it was also largely the case at the Aberfan disaster of 1966, also involving the failure of coal mine waste.

BAD NEIGHBOURS

The lovely expression 'heads I win, tails you lose' sums up a problem that is faced by developers when all the good sites are taken. If they find flat ground, it often has a flood risk, and where there isn't flat ground, they have to build on a slope. Sometimes there isn't any choice: there are no flat sites.

Then, inevitably, one or the other neighbour decides that it is time to remodel the slope. The uphill neighbour decides that he wants a flat garden, or a patio. The downhill neighbour decides that cutting into the slope, for a garage or extension, is the right thing to do. Sometimes they get away with it. Sometimes they don't. Most of this sort of work is done without any regard for how the slope was formed in the first place: there will not have been any ground investigation, and the excavation will have been done by a cheap contractor. On occasion, a slope may have been cut into to permit building in the first place, or a slope may have been flattened at the bottom to allow more light into a downstairs window of the lower property on its upslope side.

Such cases sometimes lead to ground movements of the sort that displaces a fence, and sometimes it triggers a landslide. Inevitably access is limited, and a dispute arises. It's difficult to decide who is most to blame in such cases – my thoughts on how to decide are in Chapter 10.

Bad neighbours also include other slopes that channel water towards your slope, thus causing erosion or local instability. How serious that can be depends on what the vulnerability is of elements at risk where the debris ends up.

THE SLOPE SABOTEUR

Occasionally, slopes are destabilised intentionally. I have been partly responsible for this myself, through the Selborne Controlled Slope Failure Experiment,[11] which was to cut slopes in stiff clay mirroring what a trial embankment on weak foundations does at the other end of the spectrum. A 9m-high slope was cut into a vacant length of a brick pit in Gault clay, instrumented with probably the densest array of piezometers and inclinometers ever installed, and brought to failure by pumping water down a set of recharge wells. Trials involving the deliberate failure of cut slopes are relatively few and far between, unlike trial embankments, which are comparatively common.

Trial embankments are often taken to failure, but in some cases they are not, because their purpose is to do field trials to determine the rate of consolidation or to accustom the contractor to compaction requirements.

I was once engaged as an expert in a case where a rich crook had commissioned an engineer to design elements of his new house to be built on the crest of an escarpment with rather good views over the countryside. The builder, a chum of the owner, built something radically different, and then promptly went into liquidation. Some elements of what was built 'underperformed' (as you might expect, such as odds and ends of cut slopes left in their temporary state), including a crib wall and patio that sagged dramatically and eventually failed. It transpired that the owner had bought the crib wall concrete units and had his chum build the patio, for which there was no planning permission, and which he was therefore instructed to take down.

Intriguingly, the crib wall and patio collapsed within hours of the deadline to remove it. The manufacturer of the crib units was being sued along with the engineer. However, it was obvious that the crook had taken a sledgehammer to some of the crib units in order to have a suit against the crib wall manufacturer, and the whole badly constructed edifice had collapsed around his ears. This was put to him before the court convened, and realising that the game was up, he dropped his charges and paid off everyone's costs!

At least that one was deliberate, and was uncovered, with the party pressing a complaint was unmasked as a fraud. More often than not in such cases, they get away with it.

TIPS

The nature of mining and quarrying operations is that the costs of excavating unwanted rock and soil and disposing of it are overheads, which operators wish to be disposed of at minimum cost. Wastes are therefore often loose-tipped and not compacted. If there isn't a flat site, they have, in the past, been dumped on sloping sites, as at Aberfan. Loose materials are at particular risk from slides developing into flows.

In mining and quarrying, there is always a quantity of unsuitable and unsaleable material that is extracted and which needs to be discarded. For quarries, it is the overburden: soils and weathered rock that is not saleable. What starts as an inconspicuous pile in some odd corner of the site grows, and when added to with more overburden and poor quality material that is discarded, eventually it becomes sufficiently high or encroaches on a part of the site where the subsoils are weak, and instability is inevitable.

I am always reminded of a paper where an opencast coal mine was chasing a seam down the dip, and the ever-increasing amounts of overburden were back-stowed in the earlier parts of the excavation, sliding in a big, but fortunately slow-moving, landslide towards where the seam was worked. The authors of the paper seemed unconcerned by that, and remonstrated when I had the temerity to criticise the practice. Dangerous men to be around. "*Dig faster, boyos! The landslide is catching up!*"

Mine and quarry faces, of course, always carry their own share of slope instability issues, as the extraction process may release a rock slide along a previously unknown discontinuity or discontinuities in the rock mass, and the face will, in any case, have been potentially weakened by blasting.

Some processes in mining and quarrying demand that the extracted material is crushed and processed, sometimes with noxious chemicals. The fines or tailings generated from the process are sent off to a settling lagoon. Although no doubt some lagoons have dams that will have been properly designed and constructed, many are neither, and suffer from defects such as lacking sufficient freeboard so that they overtop and erode, or become unstable because of seepage through the bund from the lagoon. If the bund fails on a big enough scale, the contents will escape as a flow, disastrous in itself, but compounded if the contents contain residues of noxious chemicals used in an extraction process.

The problem is exacerbated by the fact that intrinsically unsafe practices repeatedly emerge from the woodwork, such as 'upstream construction', where a lagoon retention bund is raised progressively by building in part on a previous stage of the bund and in part on the tailings that the bund retains. As if that wasn't bad enough, the bund itself may sometimes be constructed *from* tailings! And the authors of this abomination are astonished when it fails? I am always astonished when it does not.

Setting lagoons may also be constructed to clean water from a quarry or mine before it gets into local watercourses. My comments still apply.

Municipal solid waste (MSW) is a form of discarded material that has the added disadvantage that it releases noxious smells and attracts vermin such as rats and seagulls so that, periodically, capping materials are used to cover the surface and prevent the smells, the vermin and the discarded materials that may blow about in the wind. The wastes of the past were rather less voluminous than the present, and probably contained less discarded food, or such components of waste have already decomposed. In addition, in the UK at least, the use of open coal fires generated a lot of ash, much of which was discarded along with household waste. Prior to the segregation of wastes into recyclables and non-recyclables, MSW would have contained tin cans and glass bottles, and of course other discards like broken crockery and china, both of which give rise to sharps that can cause injuries if handled during the course of a ground investigation. Modern MSW dumps also contain a great deal of plastics: bags, babies' nappies and other things that may pose bacteriological hazards if you are unfortunate enough to have to investigate them.

MSW is often surprisingly light because of the empty containers in it as well as the basically low density of many of its constituents, such as plastics, wood and other organic materials, often attaining densities comparable with water. It is not surprising, therefore, that sometimes the sides of a filled waste pit can slide inwards in a way that you might expect with an empty pit but would not expect with a full one if unaware of the density difference.

The capping materials are often simply a layer of soil. When later buried by another phase of waste deposition, some capping layers may form preferential paths for the development of slip surfaces. When this happens, or the waste slides on a surface in the natural ground, the reinforcement *via* the content of plastics makes the sliding mass remain relatively intact and move monolithically, forming what some commentators call 'trashbergs'. The moving mass may continue to release gasses from decomposition for days after failure, and being largely methane, they are combustible.

A BELIEF IN COMPUTERS

The problem with computer analysis is that the output *looks* so authoritative, especially nowadays when the outputs are in glorious colour, that it is easy to believe that they are not only correct, but correct to the exclusion of any debate about their meaning. Let us leave out, for the moment, the idea that the software has bugs (it will have, no doubt about that, but whether or not they matter in a particular context is moot) or that the hardware may have faults, there is the problem of whether or not the right sort of analysis has been done, and whether the inputs are reasonable.

I was reminded of the case where a technician engineer failed to appreciate that a slope had ancient landslides in it, mainly because he saw no sign of them and did not look to either side of the site, and using peak strengths designed an earthfill to give horizontal gardens to a small estate of houses at the top of the slope. Understandably, perhaps, the fill was only partly built before it began to slide. The inevitable row with the contractor and the onset of winter delayed any prospect of starting works, and the following spring, things had got a whole lot worse. Expert opinions were sought, various remedial schemes were designed and criticised, instrumentation was installed in the boreholes of yet another ground investigation, and one day, I found myself in a meeting where a question on the agenda was "*Was the analysis right?*" I am sure that the questioner meant the arithmetic, but there were multiple levels of incorrectness implied.

The problem is as summarised in the Americanism, 'garbage in, garbage out'. What is put in is not just the input dataset, but a whole range of assumptions in the software, including the algorithm and constitutive laws hard-coded into it. For example, if one uses the $\phi = 0$ method, then it is easy to show that changing the water pressures in a slope has absolutely no effect. Also, if you analyse every failure in a particular soil, say for example all the cut slopes in London Clay, then you discover, unsurprisingly, that the bigger the slide the stronger was the soil. And if, for example, you have a couple of big slopes that failed soon after construction, you can develop a model that includes the strength of the clay increasing dramatically with depth but reducing drastically with time. There are elements of truth in this, but it isn't the whole picture, as it dispenses with the principle of effective stress, which, when evoked, gives cogent reasons for the finding. In fact, you probably get the results that Skempton got back in 1948![12]

Similarly, if you only have software to analyse slip circles, then you can demonstrate – because the computer can't be wrong, can it – that a thin weak layer in the soil profile has no effect on stability either. I am afraid that if slope stability is given the few lectures devoted to it in many soil mechanics courses at university, then many graduates simply won't understand this point, even after it has been explained to them.

Neither of these has anything to do with the details stored in the dataset used as input, but are errors in the appreciation of the problem parameters and some assumptions baked into the computer code used for analysis. Others are more subtle, like not appreciating that certain ways of representing the water pressures in a slope stability analysis have implications, like the hydrostatic distributions or water pressures with depth when a piezometric line is used, or the jumps in water pressure distributions from soil to soil that relate to the use of r_u and the more or less inevitable removal of zones of high and low pore water pressures from the model when r_u is used. Thankfully, the days when computer capacities were so limited that there was a risk of not taking enough slices are past, but we still get instances where the water pressures and soil parameters are simply guessed – correlations with other things making them better guesses, but still guesses all the same.[13]

Sometimes, one meets a failure of the logical process. For example, in design, it pays to take the worst groundwater conditions that one can realistically expect, as that yields a lower Factor of Safety. However, to take extreme groundwater pressures in a back analysis yields a maximum value for shear strength. On its own, that is simply misleading, but when it also persuades the analyst that drainage would be an extremely effective remedial measure, then the drains may never discharge any water, and also, they may not effect an improvement in stability if the actual groundwater pressures turn out to be low.

Occasionally, a computer analysis throws up a result that takes you by surprise. As a natural sceptic, I look first for errors in the input dataset, and then in the program. Ordinarily, with commercial software or something coded by another person, there is little you can do about programming errors, at least in the short term. Writing your software yourself, and not being arrogant about it, does lead to the computer code being looked at first. The error is usually in the dataset. I find that these surprises occasionally give me new insights, but only when I can, on reflection, explain why they arose, using simple and established principles of soil mechanics.

In a limit-equilibrium method of slices analysis, there are always approximations. The subdivision into slices, for example, if too coarse (rarely a problem with high-capacity computers) is an approximation, and a subtly different result might be obtained with a different subdivision. This also goes if you use a grid of centres, and a finite step interval in the radii specified, can also lead to subtle differences in output. I would not expect the second decimal place in an output to be exact, even in a very careful analysis – but the numbers are commonly printed out to more decimal places than that. The string of decimal places is the *precision*, but the reliable part of the answer is the *accuracy*. Don't mistake the former for the latter.

There is a further effect that the more sophisticated the analysis, the more likely it is that the result will be believed. It's an easy trap for the unwary.

BAD DESIGN

Oddly enough, the last two earth dams built in England contained elements of bad design: in the case of Kielder, a clay blanket, and in the case of Carsington, an oddly shaped core, both of which provided some very obvious paths along which a slip surface could develop, inconveniently in the former (where a toe berm prevented collapse) and disastrously in the latter, where a toe berm prevented collapse partway through construction but was not built to the size needed to prevent it when the dam reached full height.

I followed with interest the problems at the Oroville (USA) and Whaley Bridge (Todbrook Reservoir, UK) dams, where in both, the slabbing that formed the floor of the spillway was ripped out and the earthfill underneath was rapidly eroded. Sprayed or cast thin concrete facings to slopes are always vulnerable to seepage and erosion underneath, and certainly in the case of Todbrook, where the slabs were simply cast on the downstream face of the dam, it became glaringly obvious (with the benefit of hindsight, of course, which always makes the process that much easier) that seasonal cycling of the moisture content in the soil underneath would lead to the slabbing moving, perhaps cracking, but certainly becoming irregular at the joints, causing turbulence when the spillway was in use and promoting the ripping out of sections. It is abundantly clear that there are some substrates that cannot be simply slabbed over, and in any case, the slabbing needs to be structurally competent and well anchored down. The nature of joints is also important. In any case, these dams showed the erosive power of water running down a steep slope (Chapter 7).

Another example of bad design relates to drainage. Surface or subsurface water drainage of an existing landslide needs to cross the sides or toe of the slope, or even intermediate slip surfaces. Water collected over a large area must therefore not be allowed to escape from any pipework at those critical locations.

The problem is rather less readily understood when perforated pipes are installed in gravel-filled trench drains, particularly crest drains at the head of a cutting. If the collected water is run down the cutting slope in another gravel-filled trench, it sometimes seems to be a good idea to carry the same perforated pipe down the slope. It isn't a good idea: it's a bad one. The problem arises if anything obstructs the flow so that water backs up and then comes out of the pipe under pressure. That can be enough to make the gravel slide out of the trench! If the purpose of the pipe changes from water collection to water transport, it needs to be in an unperforated pipe. Blockages may arise from pipe connections, kinks of solids in the pipe itself, but other obstructions to flow may be that the water cannot get away because the pipe at the toe of the slope is already full.

BAD CONSTRUCTION

This mainly relates to embankments, but also to remedial works on landslides that proved ineffectual. A particularly common defect is the failure to compact earthfills adequately. This leads to excesses of settlement and to a more permeable embankment that would have been achieved with appropriate compaction, which is an issue with bunds on lagoons in mining and quarrying waste disposal, and therefore to the potential for failure. Railway embankments constructed in the 19th century were often just end tipped with little thought to construction for stability, and waste dumps and tips are rarely fully compacted.

Some soils do not lend themselves to compaction anyway, particularly soils with only a single particle size, and some, including some types of lightweight aggregates, should not be compacted, or the density gain destroys the point of using low-density fills in the first place.

Bad contractors may hide poor materials in an embankment, making them undetectable by covering them with something better. Sometimes the poor materials consolidate to an adequate strength – but sometimes they do not. Quarry operators sometimes discard slurry from settling ponds in their overburden tips with a similar deleterious effect.

Intense rainfall during construction can also wet up construction surfaces, leading to weaker zones if the mud is not scraped off. In any case, temporary or permanent surface water disposal arrangements can lead to erosion.

While the technical demands on site supervision are not particularly great, the supervisor must be present and must be diligent at all times.

REALLY BAD PREJUDICES

Even when you really do know better, it is difficult to shake off some prejudices that form during your undergraduate and early-career education. One of them is the idea that rather thick formations of stiff clays are uniform, isotropic and homogenous. One of my first jobs involved assisting an engineering geologist logging literally hundreds of metres (feet in those days) of cores from the lowest 50m or so of the London Clay in North Kent. It obviously wasn't all the same, but still, the profession sometimes treats the formation as though it is. Part of the problem is that we refer to *The* London Clay. It seems that the deposit might have reached a thickness of 150m or so in its sedimentary basin, but it wasn't that thick everywhere, and it didn't all start being deposited everywhere in the basin at the same time, geologically speaking. Indeed, it contains sedimentary sequences that reflect changing depositional environments through time: sea level change, distance from land (as the sea level went up and down) and even water temperatures, all of them changing the nature of the fauna, and therefore, what was being deposited on the sea bed at the time. The understanding of these processes has given rise to anew field in geological science called *sequence stratigraphy*.[14]

I remember once meeting a young engineer distressed that his ring shear tests produced an angle of shearing resistance of 30°. He brought a sample along to show me. Despite it being from the London Clay Formation, it wasn't clay, it was sand. I explained that the transitions at the bottom and top of the formation were sandy. In fact, in some parts of the formation there are sand beds associated with the sequences. It's just not 150m of all the same stuff.

NOTES

1 The film starring the actor Kevin Costner, in which the world is almost entirely covered in water except for some fabled land mass that is in everyone's dreams. Sea level rise to do this would require more ice to melt than there was locked up in the last glaciation or perhaps a hundred Antarcticas all melting simultaneously.
2 Or pretty much all year round.
3 The *thalweg* is the line of lowest elevation within a valley or watercourse.
4 As in the case of Mt Etna.
5 There is a similar difficulty between the words 'tuff' and 'tufa' when written in the field, but as with 'fill' and 'till', there are usually other clues.
6 Or for those of a more literary and less musical bent, *antidisestablishmentarianism* or perhaps *floccinaucinihilipilification*.

7 Chandler, R. J. (1979) Stability of a structure constructed in a landslide. *Proceedings of the 7th European Conference on Soil Mechanics and Foundation Engineering. Brighton*, 3, 175–182.
8 Railway practice in the UK is to call the space between the track and the start of the batter 'the cess'. They also call both fills and cuts 'embankments' (although not, apparently, on HS1).
9 Readers may already be aware that this was the landslide at Warden Point on the Isle of Sheppey, occurring in the uppermost part of the London Clay Formation.
10 The A21 Sevenoaks Bypass.
11 Cooper, M. R., Bromhead, E. N., Petley, D. J. & Grant, D. I. (1998) The Selborne cutting stability experiment. *Géotechnique*, *48* (1), 83–101.
12 He learned more about this later in his career, which is fortunate for those who follow in his footsteps – people like me, for instance.
13 In a PhD thesis in 1970, its author analysed numerous failures of cut slopes. I will save that Author's blushes by not citing the reference in full. When there was no pore water pressure information, he assumed r_u to be 0.3. It was an assumption that devalued the results. He should, at the very least, have assumed a range, and indicated a degree of uncertainty in the results. What was even less forgivable was that the slip surfaces sometimes did not accord with the geotechnical instrument readings! At the time, I was using software that expected to repeat analyses several times with different properties and water pressure assumptions, because we knew that even if the dataset passed all the tests, we didn't get the results back until the following day, and meeting the stringent requirements for formatting the data, the results probably took several days to receive. Even as far back as 1970, it was obvious that the results of a single set of assumptions could be way out.
14 For an introduction to this, consult: Coe, A. (2003) *The Sedimentary Record of Sea Level Change*. Open University Press. A great book, even if some of the later chapters are beyond me. Just disregard the chapter where it postulates that a postglacial rise in sea level flooded the Black Sea region and gave rise to the *Noah's Ark* myth. Or believe it. It's up to you.

Chapter 4

The development of ideas in slope stability analysis

NOT THE END, NOR THE BEGINNING OF THE END, BUT PERHAPS THE END OF THE BEGINNING

When you reach for your laptop or desktop computer with a software package for analysing the stability of slopes you have to realise that the analytical methods in the software have developed over a long period. The original roots lie in Coulomb's 18th-century essay[1] on earth pressures, in which he proposed that the earth behind a retaining wall with a tendency to slide down an inclined plane was only held in place by a combination of the reaction from the wall and the strength of soil acting along the inclined plane. The reaction, or earth pressure, was identical to the amount by which the forces driving the sliding *down* the inclined plane exceeded the resistances due to shear strength *along* it.

We know now, of course, that Coulomb's model for the shear strength was oversimplified, and not being couched in terms of effective stress failed to make due allowance for water in the soil mass. Observation of many failures tells us that in fact the sliding surface may be far from planar, and we know that not all retaining walls move sufficiently far for the soil to go into an active state so that the entire shear strength is mobilised. Some walls may be constructed in a way that they are called upon to resist the initial state of stress in the ground. Nevertheless, Coulomb provided a starting point. Other engineers of the time and later developed graphical solutions using force polygons to either explain mechanisms or allow some of the calculations involving trigonometric functions to be simplified.

Quite why it took so long for slopes *without* retaining walls to be analysed by what is basically the same procedure must now be something of a mystery. However, the failure of the jetty under construction for the Swedish town of Gothenburg[2] led to a development of ideas as to how to analyse the stability of slopes. Anecdotally (as I was once taught), some of the timber piles for the jetty were recovered and found to be broken off at a variety of depths. The engineer in charge plotted them on a cross section and discovered that the surface on which sliding a taken place could be represented approximately by the arc of a circle with a centre way up in the air above

DOI: 10.1201/9781003428169-4

the slope. I'm not sure that I believe the anecdote, as a published section of the landslide shows the slip surface going deeper than the pile tips. More probably, the piles were found to be rotated. Then, using a flash of insight, that engineer realised that the weight of the soil mass in the landslide caused a moment about the centre that could only be resisted by the shear strength acting round the slip surface. At the instant of failure, the driving and resisting moments had to be the same, from which he could determine the shear strength of the soil, and he could design some earthworks such that the resistances were bigger than the driving forces, and the whole system would therefore be stable.

There was, however, a difficulty in locating the centroid of the cross section as drawn and its area from which its weight might be derived. This difficulty was resolved by dividing the mass into narrow vertical strips, which simplified the calculation of the weight, and also by taking moments for all the strips, simplified the calculation of the moment. The strips might also be called 'slices'. The net result, among other things, was that not only was it possible to calculate a Factor of Safety for the failure of the soil mass along such a curved surface, but that also a range of terms entered the vocabulary of geotechnical engineers, terms such as 'the method of slices', 'the Swedish method', 'slip circles' and, of course, combinations such as 'Swedish slip circles'. For some writers, the name of the originator crept in, for the method to be known as the 'Fellenius' method'.

At this point it is worth remarking that the calculations for the single slip surface in the Gothenburg harbour case are relatively undemanding, but if the position of the slip surface on which failure was most likely was not known to begin with, there would be a problem of finding it. In the absence of any theory, that might be a matter of analysing dozens or even hundreds of 'slip circles' to find the one with the lowest Factor of Safety, and that required considerable computational effort at a time when computational aids were very primitive.

Further developments in soil mechanics theory involved the principle of effective stress, and this appears to have been incorporated into the method of slices by engineers at the US Bureau of Reclamation, or so says Prof Bishop[3] in his classic paper, where he refers to the method as being used *conventionally* at that place, which gives us the 'US Bureau Method' or sometimes 'Conventional Method' to add to the compendium of names. Water pressures were simply computed at the base of each slice along its slip surface and taken into account where appropriate in the computation of sliding resistance. The approach is sometimes also called the 'Ordinary Method of Slices', particularly in the US.

We now get to a difficult point in the development of the method, and that is whether or not to take into account the forces between the individual slices. The force from an upslope slice obviously bears on its neighbour downslope, but these are internal forces that counteract each other when taking moments about the centre. Therefore, it is argued sometimes that they can be completely ignored.[4] However, the distribution of internal forces does change the distribution of normal stresses around the slip surface from the rather oversimplified distribution that one gets if they *are* ignored. Where the soil is purely cohesive, this scarcely matters, because the shear strength is independent of the normal stress on the slip surface, and in any case, for a slip circle, all the normal forces pass through the centre of rotation and therefore have no effect on the moments. Where it *does* matter is when the soil is frictional, because there, the redistribution of normal stresses causes a redistribution of strengths, and this may in some cases mean that the overall equilibrium is affected. It also matters for other shapes of slip surface.

Prof Bishop therefore set out to develop a method that took into account those *inter-slice forces* as they had become known. He developed two methods in the classical paper I mentioned. In the simpler of the two, only the effects of the horizontal components of interslice forces were taken into account, and in the rather more complicated method, shear components were also considered. Once again, this leads to terminological complexity. Bishop referred to the horizontal interslice force method as being suitable for routine work, as the full method appeared to produce little benefit. In the paper it is derived by removing interslice shear forces from the equations of the full method. We therefore have a school of thought, largely from those taught by Bishop, in which the simpler of the two is referred to as 'Bishop's Routine Method', and alternative schools that call it 'Bishop's Modified Method', 'simplified Bishop Method', or 'Bishop's Simplified[5] Method'. Nobody, as far as I know, ever bothers with the full, rigorous or complete method! North American authors almost always use the 'Simplified' name, which is unfortunate, because in his original paper, Bishop himself refers to the method with no interslice forces at all as 'simplified' on eight occasions. So, we have to assume that no-one in that part of the world ever read the original paper or understood it!

In the conventional method, the driving forces are summed over all the slices, and so are the resisting forces. The Factor of Safety is derived from the ratio of the two sums. However, in Bishop's Routine Method, an equation was derived which had the Factor of Safety on both sides and cannot be reduced to something that may be solved directly. It must be solved by iteration. I'm sure I'm not alone in having wasted hours of my time struggling to rearrange the equation, but at least I was doing so while commuting to work on a train. It isn't possible. But, the form of Bishop's equation is such that an iterative method is indicated by the very nature of the formula. It is possible to simply guess the start point or seed value for a Factor of Safety F, feed it into the formula and derive another Factor of Safety, then take that one back into the formula, and one finds that each successive run through produces a better answer and that the process converges much monotonically to the final result. In my experience, the process is marginally faster if one starts from seed value for the Factor of Safety that is larger than the final, converged, value, but the benefits of this are slight. However, the arithmetic is several orders of magnitude more complicated than with the conventional method, making the problem of computing multiple slip circles dramatically more difficult in the pre-computer era.

Bishop found that the need to consider interslice forces at all was most serious when the slip surfaces were deep and/or there were high water pressures to consider, but that considering the interslice shear forces was not really necessary in the vast majority of cases.

THE BEGINNING OF THE MODERN ERA

It was perhaps fortunate that in Bishop's circle (no pun intended) of friends and acquaintances there were the engineer Little and mathematician Price, who were able to create a computer program[6] to carry out the necessary calculations. Their paper appeared in 1958 and marked a turning point in slope stability calculations. However, even a decade later, digital computers capable of running this kind of software were owned only by universities and the larger firms of consulting engineers. The rest of us had to wait until the early to mid-1980s and the advent of personal computers before the ability to do computerised slope stability analysis became widely available, and even then, there was a slow take-up.

In the universities, the availability of computing power was something of a blessing, as it enabled the development of more complicated mathematics without the obstacles of calculation. For example, it was now possible to solve the problems of slip circles that were not the arcs of circles, something that had long been observed. For example, in the mid 19th century, the French engineer Collin[7] wrote a book about his experiences of dealing with slides on the French canal system, and a little earlier in Britain, Gregory gave a paper at the ICE on a landslide in a railway cutting at New Cross, which prompted, in discussion, other engineers to give their experiences. While Collin attempted to fit mathematical shapes to his slip surfaces, Gregory did nothing of the sort and did not even draw attention to the shape, slip circles being something for the future, and the shape was simply what it was.

By the time of the Bishop work, however, more than a century after Collin and Gregory, it was obvious that a method of dealing with slip surfaces that weren't slip circles needed to be developed. Using methodologies that owed a lot to Bishop's mathematical treatment, Kenney at Imperial College and Janbu in Norway provided simple formulae,[8] but it was Morgenstern and Price who developed a really complicated and complete method. They were followed by several others, with the statical indeterminacy of the problem dealt with in a variety of ways, including fixing the elevation of the line of thrust of the interslice forces, fixing the ratio between the shear and normal components of interslice forces (and thus the inclination of the resultant force), and other constraints. At first, and rather widely, the methods were said to deal with 'non-circular' slips, such was the strength of the slip circle idea; later they were called 'general slip surfaces' or 'compound landslides'.

My own favourite method is due to Maksimovic[9] and is basically a thin-slice version of Morgenstern and Price's method. I contributed to it by deriving the relationships that permit the use of Morgenstern and Price's solution methodology. In particular, I like the method because it simplifies using concentrated forces, and by the time I programmed it for myself I had learnt a lot more about programming relative to my first efforts.

IT DOES NOT COMPUTE

A rather interesting contribution by Whitman and Bailey,[10] a decade after Little and Price had introduced the idea of doing stability analyses by computer, demonstrated the use of computer graphics. It was many decades later that high-resolution graphics in colour became readily available. As an aside, my own excursions into computer analysis initially employed a pen plotter to draw out the sections to scale so that I could compare them with my carefully constructed (on graph paper) ground model. Pen plotters now seem to be a thing of the past, and so I am no longer able to plot sections on tracing paper, but a laser or inkjet print can still be overlaid on the original drawing, even if it does often take a light table to see if they match. Computers now have sophisticated graphical input systems, and the light pen has been superseded by graphics tablets and more commonly by the mouse.

Whitman and Bailey showed that even in something as relatively uncomplicated as Bishop's routine (but iterative) method, there were combinations of factors which would lead to the normal effective force on the base of a particular slice or slices becoming zero or negative and that when this happened, errors were introduced into the computation. There is, firstly, the situation of a zero value for normal effective force being liable to create some instabilities in the mathematics and the rather more serious negative value reversing the direction of action of the shear strength on the affected slices. In my experience this is something of a red herring, especially when very large numbers of slices are involved so that the errors are very localised. One has to remember that in any iterative solution the answer is produced to a finite accuracy anyway, even before one takes into account that the slip surface position and shape are probably approximate, the shear strength parameters poorly defined, and the groundwater conditions inevitably contain a great deal of guesswork.

We also find that Morgenstern and Price's method does, on occasion, lead us on a merry dance trying to converge to an answer. The reason for this took a great deal of experience to explain, but the answer is very simple once the underlying principles are understood.

A theme in the work of Hutchinson[11] is that loading the head of a landslide is a bad thing to do, because not only does the load directly increase the shearing stresses around a slip surface, but if pore water pressures are generated, they prevent the increase in effective stresses that would provide an increase of shear strength to partly compensate. We would therefore have a difference in effect if the loading were to be undrained relative to what the effect might be if the load was drained. At the toe of the slope, loading would be beneficial, but more if the load were drained than if it were undrained. There had to be some position along the slip surface above which a load would have no effect. A formula was developed and tested, and this showed that the point at which a load would have no effect could be readily determined. For undrained loading, this point (which Hutchinson dubbed the *neutral point*) was where the slip surface was horizontal, and for drained loading, it occurred where the slip surface rising towards the head of the slip surface reached an angle equal to the mobilised angle of shearing resistance. For a slowly moving landslide with a Factor of Safety more or less equal to 1, this would be the angle of shearing resistance for the soil. Hutchinson suggested that finding the position of the neutral point on several sections through a large landslide complex would be of assistance in defining corridors where the landslide could be crossed on an embankment or, if necessary, in a cutting. It transpires, however, that the neutral point needs to be calculated for the desired ultimate Factor of Safety in order to resolve some little paradoxes when using Hutchinson's equation.

Now, the neutral point method is all well and good if the landslide is in a single material and its slip surface between the point where its slope is zero and the head of the landslide is a smooth curve, because then it has a very large range of possible locations. When we examine what happens while iterating towards a solution, it becomes possible to understand why some methods of analysis have that merry dance. For a particular instant in the computation there will be a location for the instantaneous position of the neutral point that, in effect, divides the landslide into two bodies, one of which has a net thrust and the other a net resistance. For the next iteration, this instantaneous neutral point may jump a long way round the slip surface and thus radically alter the ratio of size and weight between the two zones. The effect is particularly marked where the slip surface is composed of a few comparatively long, straight segments, such as may be drawn when preparing the ground model if there are few borehole locations where the slip surface has been found. The problem is particularly acute if the analysis includes wide slices with linear segments of slip surface that the Morgenstern and Price method permits and where the slip surface passes through multiple materials with very different shear strength properties. The answer is to have more curvature in the input of the slip surface. The problem is rarely if ever encountered with a slip circle and many slices because the angle change usually means that the instantaneous neutral point does not jump far in any iteration. The point was made in a paper by myself and my PhD student Hoseyni.[12]

Incidentally, Hutchinson's neutral point concept may be applied to both loadings and excavations, with the same neutral point computed for both. There is no neutral point for ground water pressures as increase of water pressures is uniformly bad wherever it occurs and uniformly good where the groundwater pressures are lowered.

THE PRESENT STATE OF PLAY

We therefore have reached a point at the time I write this at which the vast majority of stability analyses are carried out using one piece of software or another running on a laptop or desktop personal computer. The software is usually commercial, and whereas there was a great deal of choice in the past, many otherwise satisfactory computer programs did not survive the transition to a graphical user interface such as Microsoft Windows, and the choice is now comparatively limited. Different computer applications have relative strengths, and dare I say it, some have deficiencies, with developments and upgrades reducing or eliminating those deficiencies with release after release.

This brings me to the issue as to how computerised analyses are checked, reported and archived. A manual calculation (whether it is done on A4-size paper or on those old-fashioned sizes that are non-standard today) will fit in a file to be stored in a filing cabinet and will be as readable today as it was a century ago, and subject to the hazards of preserving paper, will be as readable a century in the future. Many digital storage methods are ephemeral.

I read (and review) a lot of papers on slope stability, and the ones that make my vision blur over include the papers that tell me in no uncertain terms that limit equilibrium is no more, has ceased to be, is bereft of life and rests in peace. The general alternative is to use a continuum method (for example, the finite element or finite difference method) or, indeed, a discontinuum method (such as considering the statics and dynamics of an assemblage of particles), and there is a class of papers that pushes these methods as an alternative to the 'Norwegian Blue deceased' method (with apologies to Monty Python). The problem is twofold: firstly, there are areas where other methods have applicability, and this does not mean that the limit equilibrium 'slip circle' methods do not have theirs. Secondly, the more sophisticated approaches require even more parameters than unit weight, cohesion, and angle of shearing resistance. Linear elasticity needs two more in the well-known Young's Modulus and Poisson's Ratio, plane anisotropic elasticity even more[13] and the 3D case much worse. It is difficult enough to determine the basic strength parameters, so what chance is there of getting all those extra deformation parameters even approximately correct? And it is well understood that soil is non-linear so that elastic parameters alone are not enough.

Moreover, there is a truism about any form of computation, especially when performed by computer, that 'garbage in, garbage out'. What is not always fully appreciated is that the model itself also has a bearing on the results. For example, if we use the $\phi = 0$ method, in which soils are purely cohesive, then we might predict that slip surfaces penetrate deeply, and embankments have a limiting height, whereas if we use a purely frictional model, we usually predict shallower critical surfaces and slopes where the angle is critical, not the height. And, in the $\phi = 0$ method, internal water appears to have no effect. (Its effect is, of course, encapsulated in the undrained strength.)

The underlying problem is that the outputs from computer analyses always look authoritative, especially when presented with an elaborate coloured graphic. The correct approach is to be sceptical of everything, yes, especially limit equilibrium results. And why? Because, among other things, the input parameters may be little more than a guess. Even published correlations are guesses if one takes into account only the relationships without also considering the scatter of the data on which the correlation is based. Guess in, guess out. And then there is the fact that iterative methods are solved to finite precision, slip circle analyses are done with grids of finite spacing with finite radius increments, the internal 'geology' of the slope is interpolated from a finite number of boreholes and so on.

CHECKING

At various times in the life of the personal computer, there have been reports that such and such a processor fails to compute certain results correctly, such as the infamous 'Pentium bug'. These very rare occasions are normally dealt with perfectly adequately in the firmware, but occasionally they engender a minor version of mass hysteria, something along the lines of what happened with the so-called millennium bug that was fêted to bring down civilisation when computers couldn't manage with year 2000 dates! I think that it's fairly unlikely that an engineer or geologist doing slope stability analysis on personal computers is ever likely to come across a hardware problem that affects his results, but then, there is a first time for everything I suppose. What is much more likely is that we will discover that a particular piece of software does not handle certain cases or situations very well or at all. One would discover such a situation either by the program choking on a specific dataset or giving radically different results to an alternative program for what is essentially, but can never be, identical sets of data.

Perhaps the most stupid method of checking that I ever saw was a consultant checking a contractor's temporary works design for excavation. The contractor had a computer with a slope stability analysis package and had prepared data from the ground investigation which he ran through his system. The consultant made a point of purchasing exactly the same model personal computer, obtaining a licence for exactly the same software, and demanded copies of the input data from the contractor. Having run the same data through the same program on the same model computer it then became the job of a junior staff member to compare the two sets of printouts, which, as one expected, were identical.

Proper checking requires making sure that the ground investigation information is correctly transcribed onto the cross section or sections, and that this summary is correctly transcribed into the computer. It is possible to use computer drafting systems to prepare the sections and to automatically transfer the data to slope stability analyses, which should reduce the number of errors made in transfer, but does not correct errors in building the cross section in the first place. My preferred systems of using pen plotters on tracing paper or inkjet prints overlaid on my geological cross section on a light table do help in checking that one has the geometry correctly entered but does not check such things as unit weights and soil strength properties.

Personally, I favour using different software to check the results. Inevitably, there are some differences, but they should not be large and should be within the realms of what happens with a different slice subdivision. Using different analysts and different software provides not just a check on how the dataset has been specified and input, but also between software applications.

A more difficult situation arose with one particularly well-regarded piece of software that was discovered to handle the case of partly submerged slopes incorrectly, the discovery being made after a number of flood control embankments designed with it failed. I only know if this case through a chain of 'Chinese whispers', but it could well have been the result of following a technical paper that presents the theory wrongly, such as the one by De Mello and colleagues[14] in the Skempton Memorial symposium, or it may have been the result of some computer coding errors. Such things do exist, and as far as I am aware, no software is ever perfect. The partly submerged case is poorly understood by many engineers, and those using the software were unaware that the results were misleading.

Incidentally, Morgenstern and Price derived their governing equations in a coordinate system where the y coordinate increased downwards. This may well work for mathematicians, but most engineers prefer to work in the upper right quadrant of Cartesian coordinates and make more mistakes adding and subtracting coordinates in the other three quadrants. I certainly do.

ARCHIVING

My own early forays into using computers had both programs and their data stored on punched paper tape or punched Hollerith cards. Both have their disadvantages: the cards making it easy to change programs and datasets by adding or removing cards, but being a little fragile and readily damaged in the reader or by mishandling. I once dropped well over a thousand cards down five flights of an open-tread staircase, and that caused me to be more careful thereafter. Later, I stored both cards and tapes in a steel drawer unit where they gathered dust until they were eventually no longer readable.

When I used a terminal-access, time-sharing, system, files were stored centrally. I knew this was a bad idea because periodically a new broom would sweep clean and the archives would disappear. Later, I used a minicomputer that had an 8-inch floppy drive, but the disks were fantastically expensive as well as not being readily available.

My first personal computer was an Apricot PC, now very much a museum piece. It used 3.5-inch floppy disks where the disk itself was enclosed in a plastic case with a moving shutter to protect the contents. At first, the disks were single sided, but eventually, the contents used both sides with a corresponding increase of storage capacity. At the time, the equivalent IBM personal computer used 5.25-inch floppy disks, rather like the larger 8-inch size, and there were single- and double-sided variants of that. Eventually, the next generation of personal computers for IBM used a higher-density format – but not the same as the ACT Sirius/Victor 9000 used, and when IBM launched their system with 3.5-inch disks, it used yet another format. I lost count of how many different types of personal computer floppy disks that there were, but they all had one feature in common: low capacity. But they did offer the opportunity to store data in a machine-readable form that didn't take up very much physical space.

Later on, various companies offered high-capacity storage devices. In my time I have had tape drives, Zip disks with 100-, 250- and 750-megabyte capacity, Jazz drives, and drives that would read and write CDs (compact disks) and DVDs. Inevitably, some drives would handle some capacity discs and not others, and as I write, it is becoming less and less common to have optical disc drives, especially on portable computers. I know for a fact that I was never able to read every type of disc format, and also even the ones that I could read it one time I may not necessarily be able to read today. For example, the computer on which I have written this book does not have the appropriate interfaces for the floppy disks, MiniDisks, Zip or Jazz drives, or any tapes that I still have buried away somewhere.

It is therefore the case that the storage of data on removable media is one of those ideas that sounds great in principle but, in practice, leads to the data having a very short life. Not, of course, that one can read those original decks of cards or punched paper tapes in anything other than a very specialised facility. Moreover, sometimes the data is lost with age, exposure to magnetic fields or just sunlight.

Even when the media are readable, there are two further obstacles. One is whether or not the dataset is easily readable by being stored in the digital form of plain text, or whether it is stored in a proprietary format and can only be read easily with a particular program. The advantage of the latter is that the user cannot easily modify the data so as to render it unreadable, but of course it does mean that without the correct program it is effectively unreadable anyway. It isn't just the right program, it is also the right *version* of the program. Long-lived computer applications have this tendency to change and 'improve' their data format through time.

I once looked at some results of the computer analysis from a commercial application. Contrary to what I just said above, the files from the DOS version were completely interchangeable and readable with the later Windows GUI version. Sadly, the two versions delivered different answers! This was the result of the two generations of the application using different precisions: the Windows version being tailored for a much larger-capacity computer than the DOS version and therefore using REAL variables in 8 bytes, but the DOS version only using 4. Fortunately, at an engineering level, the results were the same. It was just that the printouts showed differences that could have caused mayhem in a litigation with lawyers able to spot a difference in the results but failing to appreciate how little the effect of the third and fourth decimal places had on the engineering significance of F.

The second of the obstacles lies in understanding what the data files actually relate to, and that is a function of how the files are named, and which steps are actually preserved. It is actually rather pointless naming a file Steve123.dat to be looked at some years after Steve has left the firm, the slope has collapsed, and a bunch of experts are trawling through the data to look for causation and liability and find that there are missing numbers in the Steve . . . sequence, let alone failing to realise that those files are in any way relevant.

So much for the data, what about the results? Once again, we have to understand the developments. In my early days, the output of slope stability analysis appeared on a wide-format hard copy device that was called a line printer, because it printed a whole line at a time. Paper ran through a line printer driven by sprockets and so had removable sprocket holes, and the paper was fan folded. You could store these outputs in a big cardboard folder which used laces through the sprocket holes to keep the pages in place. The whole system was bulky and certainly did not fit into a filing cabinet in the way that manual calculation sheets did. I remember delivering quite a lot of such output to a client and, within a year, discovering that it had disappeared from their archives. On investigation, it was discovered that an engineer had seen it as a source of rough paper and had recycled it because he was writing draft reports on the back, unprinted, sides. After that, I made sure that I only printed in an area of about A4 size, so that the printouts could be guillotined down to that size and put into a ring binder of the sort that typically contained punched sheets of A4-size paper. They still didn't survive for very long.

In the early days of the personal computer, dot-matrix printers used a smaller form of a fanfold paper, and in turn, they were replaced by laser printers that utilised cut sheet paper. The smaller format of output was a blessing for laser printers, but it eventually dawned on me that the high-quality paper was expensive in the quantities that were used in the 'print everything' approach.

A question is, does the output need to be printed at all? Certainly not, as far as I'm concerned, as lists of numbers. The correct way to present the results of the stability analysis seems to me to use for preference a graphical format. And that brings us neatly to how to present the results of the analysis to let them be understood, and how to incorporate them in a report. I despair of the reports that I sometimes have to read where there is an Appendix of a hundred pages or more of stability analyses, most of which are meaningless F values for improbable slip surface mechanisms. The results are particularly useless in that form if the ground model is wrong, the soil parameters are estimated, and the ground water conditions derive from an obviously poor guess or two. Things may be worse: if, for example, the soil properties are listed in the text of a report but not listed in the results, and you have to rely on faith that the parameters in the text were used in the analyses. In combination with the other hundred pages of soil tests for 'contaminants' that are found to be absent or nearly so, it's a thick report to plough through.

But a picture isn't much good if it lacks all the data to create it, and all the parameters need to be listed in a legend. I also find it useful to draw contours of water pressure head over the slope section. This sometimes surprises people whose imagination is rather limited and don't, for example, know what a piezometric line or pore pressure ratio really implies.

REPORTING

I think that I should have entitled this section 'Data visualization, presentation, computer graphics and reporting', but that would have made for an excessively long title.

As for colour, there is a lot to be said for drawing cross sections in the same colours as are used on geological maps. In the UK, the British Geological Survey has a compendium of colours that it uses for the different geological formations, and to use different colours for sections might be a source of confusion. On the other hand, the BGS colours are rather saturated, which works well on their printed maps and also on a computer monitor, but not so well when reproduced on an inkjet or laser printer. Saturated colours produced on an inkjet tend to result in even the best grades of paper becoming wet and soggy so that the paper stretches, whereas on laser printers they give a thick and somewhat glossy layer that might peel off inside the printer. Printing less saturated or 'pastel' shades usually gives a better result.

It is always a good idea to separate reds and greens, as red-green colour blindness is more common than is sometimes realised. This does not mean the colours cannot be seen, but it does mean that they cannot be distinguished.

Monochrome printing is better in grey shades, as patterns (which Tufte[15] disparagingly calls 'chart junk') can obscure important detail.

When you have done stability analyses for a particular landslide or set of landslides it's quite easy to represent that in the form of a two-dimensional cross section with one or a few slip surfaces marked on it and the appropriate groundwater pressure information drawn on as a piezometric line, or, in particular cases, lines. The difficulty comes with slip circles, as one might have analysed hundreds of these and putting the information down on a single sheet of paper can be quite challenging. It became obvious early on when slip circles were used to hunt for a critical slip in an earthwork that in fact, many circles could be drawn from a single centre, and centres were best represented as a regular grid. The aim would be to find the worst circle for every centre and the worst centre in the grid. All well and good, you might say, if there was a unique worst circle for every centre and a unique worst centre for a very large and extensive grid. It may not be so.

Grids of centres may be specified in a variety of ways in different software applications. For example, the axes of the grid may follow the normal x and y Cartesian directions, or they may be rotated so that the grid extends further from the toe of the slope as the grid gets higher. Some applications rely on large grids, user-specified dimensions of grids, or multiple grids. I eventually settled on the idea of a regular square grid that the user could move around. Then, there is the issue of how many circles one should analyse from each centre on the grid. Options typically include all the circles passing through a given point, all the circles tangential to a given level or line, or all the circle spaced between two tangential levels. Of course, the radius increment and number of centres on the grid will have a huge bearing on the amount of computation called for. Imagine a 10×10 grid of centres with 20 radii, then potentially the rather simple mathematics (of, say, Bishop's Routine Method) becomes a much larger proposition than a single circle. The single calculation is readily done with a spreadsheet application such as Excel, but the suite of 2000 or more repetitions is best done with a custom application.

The delights of early mainframe computers, and, for that matter, early personal computers, included watching the results slowly emerge on the attached printer.[16] Nowadays, the results can be stored internally and plotted automatically. The schools of thought on plotting the results seem to be divided between annotating the grid with numbers, producing contours or isolines of Factor of Safety over the grid, and producing a colour scale map of the distribution. I dislike seeing irrational values for the contour levels, but much software seems to decide the colour map and therefore the contour interval on the basis of what produces a nice picture. If you are lucky, you may be presented with a picture showing where the worst circle lies.

Another useful 'trick' is to get the computer to plot all the slip circles with a Factor of Safety less than a prescribed value, say 1.3. If nothing else, the mass of curved lines shows you how deep an inclinometer has to be embedded for the end to be regarded as fixed. If the limit is increased progressively until slip surfaces begin to impinge on a particular asset, say a house remote from the crest of a slope, it gives you an idea what the Factor of Safety of the house, as distinct from that of the slope, might be.

WRITING YOUR OWN SLOPE STABILITY APPLICATION

Having used a particular firm's proprietary software early in my career, and found deficiencies in it, I set about writing my own. Not understanding the complex issues and resulting difficulties of this, my first efforts were aimed at the Morgenstern and Price method, which I finally got working in about 1971. In the meantime, I had actually worked through an example manually, an undertaking that amazed Morgenstern when I told him, as he had not done such an exercise. It took me rather a long time, as I started out using Algol-60 and had to switch to Fortran when the mainframe I was using suffered a fire.

Later, I wrote a slip circle program in the early 1970s and, a few years afterwards, added more and more methods to my computer analysis repertoire. I can honestly say with the benefit of experience that this was a complete and utter waste of time, and I can only recommend doing it if you are obsessive about computing and consider your own time effectively worthless.

Fortran saw me through the transition between computers and onward into the era of personal computers when I worked out that they had Fortran compilers and as much, if not more, RAM than the mainframes I was accustomed to using. My biggest shock was realising how little space my programs took up and also how fast they ran on a personal computer. I no longer had to wait in a queue to load my cards, nor fight for a terminal, and better still, I could run my stuff whenever I chose, not needing to wait until everyone had gone home so that I could occupy the computer (and the card punch) all to myself!

Just after the turn of the millennium I plucked up confidence to attempt to program something for Windows, using the Clearwin+ extension to Fortran produced by Salford Software (now Silverfrost[17]). The application was an aid to students on a residential surveying field course, replacing the DOS version that had worked for a decade but which students had begun to turn their noses up at because of its (very) dated user interface. Not only did it work, it worked well and only needed some minor tweaks.[18] That gave me confidence to produce a Windows version of my slope stability programs. It took the best part of a decade before the day arrived when I used the Windows versions instead of the DOS ones. Therefore, you will realise that

I am talking from experience when I point out what a waste it all was. You should realise that commercial software is amazingly cheap compared to the DIY solution. Fortunately, it was a labour of love, and I was learning as I went. It was also an occupation for times when there was something on the TV that I had no interest in.

There is also a huge downside to homegrown software, and that is the small user base, which means that bugs aren't so easily found, and the difficulty that one has of validating it. I once ran a test suite provided by a governmental body that included one of Whitman and Bailey's cases. Unfortunately, it is possible to misinterpret what they (Whitman and Bailey) say is the 'answer'. This misled one author[19] to the extent that at the end of an elaborate procedure that he developed, he came to some rather puzzling conclusions. I myself suspected that Whitman and Bailey's quoted results could, indeed, be misleading, perhaps confusing the minimum Factor of Safety for a slope with the Factor of Safety for a typical slip circle that appeared in one of their illustrations!

SEISMIC ANALYSES

A big advantage of modern numerical analyses is that it is – or perhaps will be – possible to take a slope through a whole earthquake virtually, although as the shock sequence for every earthquake will be different, this is likely to mean using the seismic record from well-documented earthquakes or instead using up a huge amount of computer time searching through a large set of shock patterns.

There seems to me to be two different approaches in limit equilibrium methods of slices, and one is to model the earthquake extraordinarily crudely as an outwards horizontal acceleration which is some proportion k of g, the ordinary vertical acceleration due to gravity. The proportion or *seismic coefficient*, k, has to be small, or any slope will fail if a large value is input. Some software allows the input of a seismic coefficient, and some to give different coefficients to different soils, but this ability is normally *via* an unpublished extension to the method used, which makes me question whether it should not be listed as such. If the seismic coefficient is uniform, the slope is, in effect, located in an inclined gravitational field which is the resultant of kg and g, and if your software does not permit the use of a seismic coefficient, then you can get the same result by rotating the slope geometry so as to bring the resultant of kg and g into the vertical. When you do this, however, you will see at least two important things: firstly, in an unfailed slope it may change the mechanism of failure, or the position of the critical slip surface, and the head of a slip surface may become overturned if you insist on using the critical slip surface from the non-seismic case. You should appreciate, therefore, that some numerical difficulties might arise from this second situation.

The alternative approach is to determine the critical seismic acceleration, k_c, at which F is reduced to 1. This might be by trial with different k values if your software permits, or there is a method due to Sarma that yields k_c directly. Obviously, this can be compared to any code requirement, but if you know k_c then it is possible to calculate approximately the displacements that the potential landslide will suffer for each shock with an intensity bigger than k_c. I recommend that you consult the source documents for the method, as the equation-free nature of this book does not permit a full description here.

Of course, when investigating a seismically failed slope, a back analysis variant of this approach is possible.

DEFORMATIONS AND SPEED

If you are called upon to deal with an actual landslide, its speed and runout are of academic interest, because if it was a fast landslide it is bound to have come to rest or be only creeping when you get on site. If you are on site when a landslide happens its speed is far more significant.

I have been involved with two cases where slow movements occurred that showed up in instrumentation and they later went on to be collapses – and were instrumented, so as noted previously it was possible to back-figure what the Factor of Safety had been when the observations were made. I don't think this observation is a substitute for analysis, and probably by the time that surface deformations are seen there is much you can do to arrest the movement, but in my view they should make the slope engineer tread cautiously in using EC7's mobilisation factor of 1.25 or going for a Code of Practice recommendation in the same ballpark unless the assessments of material strength and groundwater pressures are correspondingly conservative.

The problem isn't what makes landslides go fast, as if the shear resistance is less than the driving forces for any length of time, the formula $v = at$ means that the velocity will be fast. No, what counts is what makes landslides go slowly, and the absence of those factors means that they will go fast. I don't think that I would be prepared to countenance standing in the path of a landslide having computed that its runout wouldn't reach me, and I recommend that others show the same instinct for self-preservation.

The factors that make landslides move slowly include that the geometry is changing, or that the landslide meets some obstacle in its path. Internal factors such as shearing making the soil dilate so that pore pressures reduce may help slow clay landslides, but the corollary, which is the soil contracting and pore pressures increasing must have the opposite effect, that is to speed a landslide up. It is one of the numerous reasons that we like fills to be well-compacted, which means they will dilate when sheared and not contract, as do loosely dumped mine and quarry waste. In extreme situations where the soil contracts it may be enough to turn a slide into a flow, and flows are usually on the fast end of the landslide velocity scale.

STABILITY CHARTS

Taylor, as long ago as the 1930s, recognised that there would be difficulties in computation, and published a set of charts from which the Factor of Safety of a slope could be determined. His charts only work for simple geometries, and he prepared a chart specifically for the $\phi = 0$ case as well as for soils exhibiting c-ϕ behaviour, although his charts lack anything to do with water pressure. The charts have been reproduced innumerable times in soil mechanics textbooks, and frankly, the limitations of simple geometry and total stress mean that I have had very little use for them. However, at the time of the Carsington Dam failure, it was noted that with the shape of the core, which almost allowed a slip circle to develop entirely in core clay, and with the height and slope of the dam, the target undrained strength of the core materials (60 kPa) doomed the dam to fail, and this could be ascertained in a few minutes using those old charts!

Having access to a computer meant that Bishop and Morgenstern could also produce charts, but this time using effective stress properties and taking into account the internal water pressures, although in a simplified way. The range of properties that they chose was found to be insufficient in practice, and so the charts were extended once by O'Connor and Mitchell[20] and later even further by Prof Chandler of Imperial College. Cousins had a go that you may see in some US textbooks, and so too did Spencer, the latter using his method that was a theoretical advance on Bishop's Routine Method.[21] As is the way of such things, Morgenstern re-entered the fray and produced a set of charts for slopes subject to rapid drawdown. No doubt there are other such solutions that I have missed, but a personal computer and a good bit of software beats the charts hands down except for in particular cases such as the Carsington Dam case I already mentioned, and so I won't bore you with the respective references.

THE DOWNSIDE OF COMPUTERS

I find this difficult to write, because I am an enthusiastic user of computers, and have been since I understood what they could do. Understanding what they couldn't took a lot longer! The downside is that they convince people that their results are right – there is a religion about computers that they have the ability to take the *sin* of guessed data and turn it into the *righteousness* of their answers.

The other downside is that it allows the development of increasingly more complicated theories that it seems one will never have the underlying parameters to use, thereby increasing the number of times that guesswork comes into play. Most theories developed in recent decades demonstrate the general effect, and in some papers the authors produce formulae and ideas (like the De Mello paper) that are demonstrably wrong. Being published, I am afraid, does not mean being right, and being aggressive when you are wrong and found out just makes people know that you are unpleasant, as well as wrong.

NOTES

1 The colophon from Coulomb's essay of 1773 on earth pressures has been used on the cover of Géotechnique from the very beginning, and an accessible review of the man and his work appeared in Geotechnique: Golder, H. Q. (1948) Coulomb and earth pressure. *Géotechnique*, *1* (1), 66–71.
2 A good review of the Gothenburg Harbour landslide case is in: Petterson, K. E. (1955) The early history of circular sliding surfaces. *Géotechnique*, *5* (4), 275–296. The paper is probably more accessible than the original publications.
3 Bishop, A. W. (1955) The use of the slip circle in the stability analysis of slopes. *Géotechnique*, *5* (1), 7–17. A particular difficulty with citing this paper is that the canonical version appeared in *Géotechnique*, but the paper appeared in a Norwegian Geotechnical Institute special publication the year before, so than one has the choice of 1954 or 1955 for the date of publication.
4 And they can be, if the soil strength is purely cohesive, or '$\phi = 0$', which gives its name to yet another variant of the method of slices.
5 Bishop didn't call his iterative method 'simplified'. In fact, he calls the variant without any interslice forces at all 'simplified'. Hence, it is wrong to attribute something to Bishop when he meant and described something else by the term!
6 Little, A. L. & Price, V. E. (1958) The use of an electronic computer for slope stability analysis. *Géotechnique*, *8* (3), 113–116. Bishop proudly asserted to his students that at the time, this was the most complex software running on a computer in Britain. It probably was – at least as far as Bishop knew, as he was probably not aware of the Bletchley Park codebreakers.
7 Collin, A. (1846) *Recherchés experimentales sur les glissements spontanes des terrains argileux*. The English translation by W. R. Schriever and published by the University of Toronto Press in 1956. It includes a foreword by Skempton which reads oddly today in the light of developments in knowledge since then.
8 Kenney had a go while a student at Imperial College: Kenney, T. C. (1956) An examination of the methods of calculating the stability of slopes. MSc Thesis, Imperial College. Janbu also adopted a variant of Bishop's method(s) before eventually coming up with what he called the *Generalised Procedure of Slices*.
9 Maksimovic, M. supplied me with its underlying theory when he was packing up after his MSc studies in 1971, and I was just embarking on mine. I subsequently worked on the solution method. It is presented in my book, the *Stability of Slopes*. A great benefit of the method is how easy it is to include forces such as anchor loads. You may also find a treatment there of Morgenstern and Price's method, which saves me giving references here!
10 Whitman, R. V. & Bailey, W. A. (1967) The use of computers for slope stability analysis. *ASCE Journal of the Soil Mechanics and Foundations Division*, *93* (SM4), 475–498.

11 Hutchinson, J. N. (1977) Assessment of the effectiveness of corrective measures in relation to geological conditions and types of slope movement. *Bulletin International Association of Engineering Geology*, *16* (1), 131–155.
12 Hoseyni, S., Torii, N. & Bromhead, E. N. (2011) Residual strength measurements for some British clays. *Landslide Science & Practice*, *1*. Proc. 2nd World Landslide Forum, Rome. October 2011 (Editors Margottini *et al.*, Published by Springer-Verlag, Heidelberg, 203–209).
13 Five, if the axes of anisotropy is horizontal-vertical, six if there is a rotation. See for example, Pickering, D. J. (1970) Anistotropic elastic parameters for soil. *Géotechnique*, *20* (3), 271–276. The brevity of the paper conceals its complexity.
14 De Mello, V. F. B., Larocca, E. S., Quintanilha, R. & de Meireles, E. B. (2004) *Reappraising Historical Coincidences That Radically Misled Slope Destabilization Analyses of Homogeneous Earth Dams.* Advances in Geotechnical Engineering: Skempton Memorial Symposium, ICE. Unfortunately, this paper is not an advance, because the theory it presents is completely wrong. I suspect that it received only cursory review.
15 Tufte, E. R. (2001) *The Visual Display of Quantitative Information.* Published by its Author.
16 Line printers were noisy as they hammered out a whole line at a time. Early dot-matrix printers gave a completely different sound as the printhead moved from side to side. Neither was anywhere near as hypnotic as watching a pen plotter whizzing across the page. Modern desktop printers are simply annoying, partly because no matter how much money they cost and how fast they seem initially, when they have been used for some time they do seem desperately slow and fantastically expensive to run. Their paper jams are more than just annoying.
17 I'm still very much attached to this software, see: www.silverfrost.com, where you can find my treatise on programming for Windows with this amazing software.
18 Like not letting students specify a 1:1 scale for plotting, which generated lots of blank pages of output!
19 Bell, J. M. (1968) General slope stability and analysis. *ASCE Journal of the Soil Mechanics and Foundations Division*, *94* (SM6), 1253–1270.
20 O'Connor, M. J. & Mitchell, R. J. (2011) An extension of the Bishop and Morgenstern slope stability charts. *Canadian Geotechnical Journal*, *14* (1), 144–151.
21 Or it would have been had his formulae not contained an error in his moment equation that becomes significant for deeper surfaces. His method requires hundreds more calculations than does Bishop's Routine Method.

Chapter 5

Modern methods and back analysis

LIMIT EQUILIBRIUM MAY NOT BE THE ANSWER IN THE FUTURE

The classic approach works, in most cases, well enough, and yet it is a rather simple and sometimes brute-force method. How can that be – that it works? It doesn't tell us anything about strains or deformations, certainly. That should be a deficiency, but it does not appear to be. The reason is that in limit equilibrium and the method of slices we have both things needed for a plasticity solution: a stress field in equilibrium and a mechanism or 'compatibility'. In plasticity, the former gives us an upper bound to the true solution, and the latter, a lower bound. If the two match up, then we have the 'right' answer. In the slope stability analysis methods of the previous chapter, we have an approximate upper bound and an approximate lower bound at the same time, and thus we have an approximately right answer – as right as can be, given the inputs. It is always important to remember that all calculations are approximate, and the analysis is at least as good (or bad) as our estimates of strengths and water pressures! It cannot be better. And it still does not tell us anything about strains or displacements.

Throughout geotechnical engineering, there is this problem. The bearing capacity of foundations is (usually) calculated by a limit equilibrium method. Earth pressures on retaining walls are (routinely) calculated by a limit equilibrium method. Deformations are calculated separately, sometimes using elasticity solutions, and sometimes using consolidation theory. Naturally, some people don't like this and want to use methods that *do* involve strains and displacements as well as simply calculating the reserve against collapse. In structural analysis, for example, the methods in common use *do* allow the calculation of strains and displacements, and the plastic analyses are viewed as overly complex and difficult to understand.[1] There is therefore a major difference between the approaches in these two fields of civil engineering: geotechnics and structures.

DOI: 10.1201/9781003428169-5

CONTINUUM METHODS

A continuum method uses the stress-strain behaviour of the rock or soil to predict how loads including self-weight cause deformations and strains within the object being analysed, in this case, a slope. Continuum methods fall into four main groups: finite differences, finite elements, discrete elements and boundary elements. Finite difference methods have the longest pedigree and may be thought of as being approximate solutions to the governing partial differential equations governing soil and rock behaviour, with finite element methods solving the corresponding integral equations. Using an equally oversimplified explanation, discrete element methods treat the soil or rock mass as a set of large particles which interact with their neighbours, and boundary element methods evaluate what happens inside the slope based on what happens at its surface and with some idea of depth-related effects.

In my experience, which includes programming some (but not all) of the approaches as well as using them, finite difference methods are easiest to understand but difficult to implement except for regular grids of nodes in problems with rectangular boundaries, whereas finite element approaches lend themselves to irregular shapes of the boundaries and internal subdivision of the problem into soil or rock zones.

When using continuum methods, it is highly unlikely that you will have programmed anything from scratch, but instead, you will be using commercial, or at least, proprietary, software and that means that you may be straightjacketed into using the facilities in the program or the way in which you use it, so that experience with one computer application does not necessarily mean that you will be equally as adept using another. The learning curves are correspondingly longer when switching from one program to another than they are with limit equilibrium programs that, after all, are much simpler and so inevitably have generic similarities. In view of the complexities, my advice is to receive training in using any application long before you have to use it in earnest.

The use of continuum methods advances along with the increases in power and capacity in computing technology. When Little and Price published the paper referred to in the previous chapter in 1958, their program stretched the abilities of a mainframe computer to its limits. A decade later, most large consulting organisations and many universities had computers that could easily handle the job, but mostly using home-grown software. The next decade saw the growth of terminal access, and home-grown was largely replaced with commercial software. Another decade after that, just about any organisation even down to a one-man band could afford a personal computer that could make the calculations with ease, and a different type of commercial software became common. Home-grown is now the exception rather than the 1960s rule.

A similar process has occurred with continuum methods, but about a decade or decade and a half behind. Procedures that required overnight runs on a fast mainframe computer now solve almost instantly on a personal computer, and non-linear procedures get faster all the time. I remember seeing a cartoon in which an engineer wielded a pocket calculator with a button labelled 'FE Analysis'. In truth, finite element analysis could probably be carried out on a cellular phone today if someone had written the application.

As far as I can see there are still major drawbacks to the approach of using a method like finite elements, and that is in the availability of parameters not only to describe the shear strength of soils, but also how they deform throughout the stress-strain relationship. That problem is not all, because if the analysis is coupled, that is, the stress changes result in water pressure changes, which also dissipate, then appropriate parameters are also needed for that aspect of behaviour too. Add to that the necessity of following the correct stress-path, if the deformations are to be meaningful, and that means that tests to determine properties need to follow the same stress-path as in the field, and the difficulties are further compounded. There are some cases in which the answer might just be independent of the stress path, but then why add layers of complexity to the solution, any of which might just degrade its accuracy?

Finally (perhaps), there are limitations in the method itself: for example, elements that use integration points in certain patterns impose a sort of stratigraphy on the problem that may or may not be there, and some elements respond to having limited distortion better than being highly contorted.

THE FINITE ELEMENT METHOD

The finite element method is the approach that I have used most myself. A finite element mesh consists of a set of triangles or quadrilaterals called *elements* that connect with each other at certain points called *nodes*. The soil or rock deformation behaviour inside each element is assumed to have a simplified behaviour relative to reality.

The number of nodes per element is variable, being more if the element behaviour is complicated, or fewer if the element behaviour is simple. In the simplest case, nodes are only present at the corners or apices of the element and the sides are straight. Complicated elements may have curved sides with nodes along them or even sometimes inside the element. In the commonest stress analysis formulations, displacements are computed at nodes, with stresses and strains at internal points, and it is at those internal points that approaches to material failure are computed.

A half-century ago, every detail of a mesh had to be specified by hand, but modern software can generate a fine mesh automatically. Even so, some aspects of the results obtained depend on the subdivisions in the mesh. For some classes of problem, the boundaries at the bottom and sides of the mesh need to be remote if deformations are to be computed reliably, and thus even in a non-linear solution, large parts of the mesh really contribute very little!

In addition, there are classes of problem for which the solution is not significantly improved with a fine mesh, for example in the vicinity of a corner in an impermeable boundary to a seepage problem, which is a singular point. Another issue I found when modelling the dissipation of pore pressures was the 'jumpiness' of the solution at early times due to the rapidity of change and the inability of different types of element to handle it correctly.

COUPLED AND UNCOUPLED SOLUTIONS

A 'halfway house' is to use a continuum method to analyse seepage through a slope, and then to use the water pressures determined in that analysis in a limit equilibrium–based analysis. However, in anything that is not 'geology free' soil mechanics (*i.e.* a single, uniform, homogeneous soil mass that obeys the simplest laws of soil mechanics to the letter) then it is important to understand the different zones that make up the slope, and the possible variations in hydraulic conductivity and boundary conditions. The upshot is that the determination of Factor of Safety then becomes a range, not a single value. You would be wise to also see the other material properties as ranges rather than single values.

So much for an uncoupled solution: a fully coupled solution will work out the internal water pressures along with the stresses. Then, there is a matter of the time dimension to take into account as well as everything else.

It is important to remember that every type of its analysis in geotechnical engineering only ever produces approximate results in respect of reality no matter how good they are at modelling the situation presented for analysis. Continuum methods have the advantage that they do at least produce some estimate of deformations at the surface and interior of the model, whereas the limit equilibrium approach does not. On the other hand, limit equilibrium does permit us to estimate the Factor of Safety of failure mechanisms other than the most probable one, as we just select slip surfaces of the appropriate shape, whereas the continuum method runs out of steam once it is shown that most probable failure mode.

Some of these difficulties with all kinds of analyses can be resolved by analysing field behaviour. This is called 'back analysis', and I will describe it as applied through limit equilibrium analysis methods to determine shear strength. It does have other applications, but for brevity, I will leave those to your imagination. I will develop the ideas in the context of limit equilibrium, then extend them to the continuum methods.

BACK ANALYSIS

I have found it difficult to write this section without relying heavily on a paper written for the 2012 Banff (Canada) International Symposium on Landslides co-authored by my research student Mehdi Hoseyni, and therefore making great use of traditional, Harvard-style (Author:date) references rather than the endnote style I have used elsewhere. References not in the endnotes are therefore listed as if they constitute one final endnote.

Customarily, limit equilibrium stability analyses take a variety of inputs: the slope geometry, the groundwater pressures, the shear strengths (usually as parameters so that the actual strengths are a function of the first two inputs) and, if necessary, some seismic inputs, ultimately providing some index of stability. A back analysis starts from knowing the value of that index of stability, usually because the slope failed or the landslide continued moving, and then the analysis allows us to figure out one or other of the factors that would ordinarily be an input. Most commonly, it is the shear strength, but it could be something else such as the pore water pressures causing failure, or the seismic shake.

Back analyses[2] are routinely undertaken in slope stability investigations and in other geotechnical engineering fields, especially to determine strength parameters, because it is known that there are serious limitations with laboratory determination of shear strength. In the literature, different authors write with satisfaction about the results from their analyses, and another group writes with dissatisfaction (or cautions in the use of the method). There are clear reasons behind these different attitudes to the back-analysis approach, and equally clear scenarios where satisfaction is likely or unlikely. Sometimes the problem results from lack of clarity as to what a back analysis can actually do, or how good laboratory tests are. The whole point is that analysing a landslide solves some of the problems associated with sample size and therefore the representativeness of the test to the landslide problem. Indeed, if the design of remedial measures for a landslide is based on a back analysis, then many of the inaccuracies in geometry and groundwater become of lesser consequence.

Back analysis by a limit equilibrium method can only indicate the value of one unknown at a time, and therefore it really cannot be used, at least, not easily, if more than one soil type is present, although there are ways round this. One of the ways is to back analyse a collection of cases in the same material, ideally spanning a range of landslide sizes, and therefore a range of effective stresses.

When applied to a single case history which is possibly unique, the inaccuracies in the analysis predominate and make the interpretation less clear, although it does provide a useful foundation for the design of remedial measures, as many of the unknowns cancel out between the back analysis and the more conventional prediction of the margin of safety for remedial works.

Back analyses of a landslide or landslides require deliberation on the following factors, primarily those relating to the overall external and internal geometry and the soil properties, apart from the strength which we are setting out to measure. By geometry I mean the shapes of the topographic surface (in 2D or 3D) and the shape of the sliding surface (in 2D or 3D), together with the answer to the question "*Is it a matter of one material or several?*"

To determine soil properties in terms of effective stress you need the pore water pressure distribution along the slip surface and, if the failure took place in a partly submerged slope, the external water level. The soil unit weight or density[3] is also important, but normally this lies within quite a small range. Finally, you have to decide on which method and which software to use, and to be certain that the Factor of Safety corresponding to the case you wish to analyse is actually 1, or else you are finding a figure for a strength parameter that is only a bound to the 'right answer'.

For example, if you analyse with zero water pressures, you get a value for shear strength that it is probable is less than the true value, or analysing with a piezometric line at ground level must provide an upper bound to that true value – all other things being equal, of course, and the water pressures not being artesian.

While most modern limit equilibrium computer codes are found to give closely similar results, there may be slight differences between the methods and programs. Computers are occasionally found to have faults, even down to design errors in the CPU,[4] but for practical cases such errors can usually be ignored. Some of the factors above, if misinterpreted, will clearly lead to the 'wrong answer', as pointed out on several occasions.[5] In the method of slices, the distance between ground level and slip surface is a critical factor in determining stresses, and errors in topography lead to errors throughout the analysis related directly to depth.

So, let us consider these factors one by one.

GEOMETRIC ISSUES, EXTERNAL AND INTERNAL

There are at least three elements to this issue: (a) the shape of the topographic surface, (b) the shape of the slip surface, and (c) the shape of any internal interfaces between materials if there are more than one soil or rock types present. The orientation of any cross section(s) chosen for analysis relative to movement is critical in 2D analysis, as the section(s) have to align with the directions of movement, and this may require dog-leg sections, or sections that do not appear to be normal to the contours. Two-dimensional analyses are usually done on a 'principal cross section', for which the position is chosen in the middle of a slide, and where it is orientated as far as possible in the direction of movement (sometimes normal to contours, but not always). Less commonly, multiple sections or 3D analysis are chosen. Clearly, the topography on the principal cross section(s) needs to be representative of the slide as a whole, or the analysis in turn will not be representative.

Since the shape of the topographic surface is usually defined from terrestrial or aerial surveys with ground verification,[6] the post-failure topography should contain few errors. Reconstruction of the topography pre-failure can sometimes be approximate, especially for natural slopes with ancient slides, as they do not have the regular geometry of a cut slope. However, the depression at the head of a slide should match, more or less, the bulge at the toe, once you have allowed for a small amount of bulking, and this can allow a check on any assumption. There must also be some influence from the effect of the shape in plan on the three-dimensional stability of the slope. An inward curvature provides lateral support but may constrain outward seepage and thus increase pore water pressures, resulting in lower stability. Conversely, an outward curvature may set off decreased lateral support with improved drainage of the sides of the landslide. While a considerable literature exists on 3D methods, the number of useful case records is much smaller.

SLIP SURFACE – OVERALL SHAPE IN 2D

The shape of the sliding surface is often determined from visual identification of slip surfaces in samples recovered from boreholes which penetrate into solid ground underneath the landslide. In very shallow landslides, the visual identification was, in the past, sometimes done in the walls of trial pits, although the support required for safe access nowadays means that most of the soil is obscured. Much more rarely, visual identification in-situ of slip surfaces is done in shafts and large-diameter drill holes. Some instrumentation such as inclinometers, blockages in piezometer tubes, *etc*., provides indications of where the borehole crosses the slip surface.[7] However, instruments take time to register the movements, and at least visual identification is almost instant. Sadly, most techniques of drilling recover only part of the geological record, and the missing part can easily be the location of the slip surface, besides which, slip surfaces are difficult to identify accurately, if at all, and looking for them requires that the cores are available for inspection, during which process they are destroyed. Commonly, there is very little positive identification for any landslide. When possible, multiple boreholes at any location make it possible, in principle, to determine the local slope of a slip surface as well as its position. This can be done sometimes by measuring the inclination of the slip surface in the core. When not only the position but also the local inclination of the slip surface has been determined, it is rather better than just having a single fix on position.[8]

The number of good records[9] of slip surface shape that have been determined over a significant part of the slip surface is much smaller than the number where it has been fixed with a few points. Cost pressures also inevitably lead to having fewer boreholes and other exploratory excavations than is desirable.

In a fresh landslide, the position of the head scarp clearly defines the extreme position of the slip surface, although as a head scarp degrades, the initial breakout position and slope are lost, and further failures may eventually conceal the slip surface outcrop altogether. In contrast, the breakout of the slip surface at the toe of a slide is almost always concealed in a thrust zone unless the slip surface breaks out in the face of a slope[10] (*i.e.* it is 'perched'), in which case toe debris falls away and leaves the feature exposed.

Even where the visual identification of slip surface positions has not been possible, then the use of inclinometer instruments may permit the identification of these positions to be made after – sometimes long after – the subsurface investigation has been completed. This relies on there being sufficient movement taking place to distort the inclinometer access tube. Deschamps & Yankey (2006) show 5 inclinometer locations for the Grandview Lake Dam in the US – usually there are fewer, but at Selborne in a research investigation there were twice as many. In the Carsington failure, a request for an inclinometer was turned down by the powers that were, although some USBR-type settlement gauges had tubes that were kinked, and that should have been warning enough for anyone. If the movements are relatively rapid, then the inclinometer access tubing may shear off in between measurements before the position of the slip surface has been adequately defined – although an inclinometer, or any other tube for that matter, can be plumbed to find the depth of blockage.

As in the case of investigatory boreholes that have not been located precisely along the principal cross section, it may be necessary to project the slip surface on to the section line. 3D analysis, in principle, resolves the problem, but it is rare for enough information to be available.[11] In time, the topography of an active landslide may vary considerably, but the slip surface position remains largely unchanged.

Where significant proportions of the slip surface are controlled in position by a weak bed then it is necessary only to identify this position by direct visual or instrumental means in a few of the boreholes (or at outcrop) as the existence of the slip surface can be reasonably inferred from the stratigraphic succession. It is then difficult to define precisely the curved rising part of the slip surface between the bedding-controlled basal shear and the head of a compound landslide. The shape of this part of the slip surface has a great bearing on the magnitude of the active thrust that drives the landslide, but it occupies only a small proportion of the footprint of the landslide. Consequently, its effect and significance in a back analysis needs to be determined by repeated trial analyses. The accuracy of a ground model may be compromised if it is not possible to locate boreholes and pits precisely along the principal cross section. It is then necessary to decide whether to project boreholes simply at right angles to the principal cross section, or whether there is sufficient information on the geological structure to project this information along the strike. Projecting along the strike often appears rather arbitrary despite its being, in principle, more accurate, because the ground levels for the boreholes are not the ground levels on the cross section. A variety of factors may lead to this inability to locate boreholes precisely along the principal cross section. An example is where inclinometers of great depth need to be installed. The weight of the inclinometer cable requires vehicular access to the top of the borehole, or perhaps for reasons of land ownership and permission for access, boreholes simply cannot be installed in the most desirable positions.

SLIP SURFACE AND TOPOGRAPHY – 3D SHAPE

It is commonly stated in the literature that neglect of end effects – *i.e.* the transverse curvature at the sides of a slip surface (or 'ends' if one considers the slip to be akin to a cylinder) – raise the Factor of Safety, and thus lower the back-calculated shear strength by various estimates from as little as 5% to as much as 30%. However, almost always such advice considers that the piezometric line and soil properties are similar for these 'ends' as in the main body of the slip. If the ends do not have the same properties, the conclusion can be misleading. As in nature, bowl-shaped and transversely curved slip surfaces are common; then we must either consider our assumptions to be mistaken, or for nature to operate unconservatively.

To a large extent errors associated with the representativeness of the principal cross section and in the movement direction are removed if the back analysis is done in 3D. There are 3D methods out there with varying degrees of sophistication, but I have found that an extension to Bishop's Routine analysis[12] works well for me, although the data preparation is so laborious that I developed a hybrid version of it based, in part, on a finite element style method of dividing the whole mass into columns that are curvilinear quadrilaterals in plan.[13] I don't think I would ever search for a critical slip surface in the way I might for slip circles and only use 3D for back analyses and then only in special cases.

Nature, it seems, always gets it wrong – according to soil mechanics theory, or so say rather a lot of authors. The problem is that if you consider a slip circle to actually be a slip-cylinder, then adding 'ends' to it always seems to make the Factor of Safety bigger, as if this is some sort of a problem. In the literature, people add plane ends, or hemispherical ends or conical ends. All it means is that in the cases where there is the problem of ends adding to the Factor of Safety, then the 2D Factor of Safety has to be less than 1 for failure to occur. There must be cases where the failure obviously has to be 3D – one of them is the case of localised loading that applies when you have a heavy piling rig on a thick piling platform, although it is a toss of a coin whether that is slope instability or bearing capacity theory that governs the answer. I have pondered this issue, and no doubt there are influences of the topography, the internal geometry of strata and the distribution of water pressures to take into account, as if 3D analysis isn't more complicated than 2D without considering anything other than 'geology-free' soil mechanics in the first place.

As well as the 3D shape of a slip surface, there is also the 3D shape of the topography to consider. In one case, an excavation made in a river bank released a large pre-existing ancient slide that was unrecognised before it moved. The footprint of the landslide was larger than that of the excavation. The principal cross section included the excavation, and thus would have been unrepresentative. Clearly, only 3D analysis can handle this case. Similar localised excavation may be the result of gully formation in the toe of a landslide, or irregular coastal retreat. Localised loading on the head of a pre-existing landslide may also cause the whole to move.

In the absence of an effective 3D analysis, there is the suggestion that the 3D effects could be accommodated by taking a weighted average of results from different 2D sections through a landslide. Presumably taking more sections improves the accuracy, rather like taking more slices.

SLIP SURFACE AND A TENSION CRACK

Various authors raise the issue of tension cracks in analysis and point out that in soils with high cohesive strength components, a water-filled tension crack in the analysis can influence the results. Where the soil is largely frictional, the calculated thrust in a tension crack may not be significantly different from the calculated active thrust from the uppermost part of the slip surface. Water thrusts in joints can seriously affect the stability of rock slopes. Generally, however, I am not a great enthusiast for water-filled tension cracks as a mechanism for failure, as in most cases the existence of a deep enough tension crack probably indicates that a slope will fail soon anyway even without a water thrust.

When I looked through my records of projects where a slope that eventually failed had begun to show displacements, and the necessary information was available to analyse that state in comparison to the conditions at failure, I realised that instruments would pick up deformations in slopes where the Factor of Safety was a conventionally acceptable 1.3, but they would only be seen in the instrument records if they were examined very carefully. When the Factor of Safety sank to 1.2 and below, the instrument records showed very clearly what was happening. By the time a Factor of Safety of 1.1 was reached, the deformations were clearly visible in the ground surface. This leads me to believe that a slope exhibiting a substantial tension crack probably is not far from collapse, or has been in that state at some time in the past. Of course, such things as a cessation of heavy rainfall may simply have saved the slope at the eleventh hour, and the Factor of Safety when the slope is inspected may be higher than my rule of thumb indicates, but it has certainly been close to the brink even if it is not there currently.

PIEZOMETRIC CONDITIONS

For back analysis of a single slip surface, the 'piezometric line' method of defining pore pressures is usually adequate. Clearly, the position of this must be based on field data: if it is not, the result could be worthless. Using the pore pressure ratio r_u tends to damp out both highs and lows in the pore pressures, so that this method is usually a poor choice.

In 2D, it is important for the piezometers to be located close to the principal cross section(s), or the projection method might influence the result. Now, there is no geological strike to provide an indication of the way to do this projection. Instead, the projection must be done at right angles to the direction of flow. It will only be an extremely rare circumstance where this is possible with any certainty. Worse still, piezometers may have to be located outside the active area of slipping, with some doubt as to whether pore pressures are the same inside as they are outside of the slip. Piezometers are often destroyed by slide movements. Similarly, some parts of a landslide may simply be inaccessible, *e.g.* in the Beacon Hill landslide at Herne Bay (UK), nearly half of the landslide was situated under the beach and therefore under the sea. With hindsight, the pore pressures in this area were probably overestimated, as a hydrostatic distribution of pore pressures under the sea was assumed, and later investigations at Sheppey (UK) in the foreshore showed extremely low water pressures in the equivalent area due to erosion and undrained unloading.

It is inevitable that pore pressure information is incomplete, and some assumptions need to be made, if only to fill in between the measurements. In the back analysis of the Sheppey landslides, the pore pressures were found to be strongly influenced by undrained unloading. It was then possible to reconstruct approximately the pore pressures by reference to the amount of undrained unloading that had occurred. This was done both for sections dated before the instrumentation was installed, and also after it had ceased to function as a result of the activity of the landslide.

Sometimes, pore pressures are controlled by a nearby permeable bed which provides limits to pore pressure.

An overestimate of the pore water pressures (*i.e.* piezometric elevation) underestimates the true Factor of Safety F, and this reflects in an overestimate of the shear strength in back-analysis and *vice versa*.

SOIL DENSITY, SHEAR STRENGTH, AND ZONATION

The density of soil is rather easy to measure. However, in many site investigations this property is not a priority. Why this should be results from the common observation that many soils are approximately twice as dense as water, and the exceptions are normally extremely easy to detect, for example in the case of peat. Indeed, for some geotechnical purposes the density of soil is not particularly important. It is important in the stability of slopes. Where soils are lighter than the notional twice the density of water, the sensitivity to pore water pressures is increased and *vice versa*, and this reflects in the back analyses. Where the soil is 'frictional' in character (*i.e.* c' is negligible) and devoid of water pressures, the soil density for a single soil case is immaterial.

Rocks are often denser than soils, as they usually have a much lower porosity. Inevitably, the increased density makes rock slopes marginally less susceptible than soil to the influence of internal water pressures.

A standard assumption of limit equilibrium analysis is that there is a uniform mobilisation of shear strength around the entirety of the slip surface. This assumption is reasonably well justified if the landslide is moving slowly. Such an assumption does not mean that the shear strength or even the shear strength parameters are uniform around the whole slip surface. Consider the case of a bedding-controlled compound landslide with a significant proportion of the slip surface running along a weak bed in the geological sequence or an earthfill on a weaker foundation. In such a case it is likely that the residual shear strength parameters operative along the weak bed are less than the parameters operative along the rising curved section of the slip surface as it approaches the head of the landslide. The difference between these two sectors may be small, for example in a comparatively uniform deposit where the weak bed is only slightly weaker than the surrounding material, or it may be large. An example where the difference is large comes from Barton on Sea, where the basal bedding-controlled shear surface is formed in plastic clays, and the heel zone developed through sandy gravels.

IS *THE FACTOR OF SAFETY* ACTUALLY IS EQUAL TO 1?

Occasionally, this problem presents itself. In a first-time failure, the onset of detectable movements occurs with lower pore pressures than collapse; as at the Carsington Dam. At the instant of collapse, instrumentation ceases to operate in most cases, even if it is present in the first place. Further complexities result from non-uniform mobilisation of shear strength as parts of the slip surface pass peak and the strength decreases to residual at different rates.

Clearly when a landslide occurs as a series of failures it is wrong and misleading to analyse it as though the whole failure occurred all at once. Some landslides 'creep' because of irregular and changing pore pressures in the landslide body or because slow erosion at its toe occurs. In the former case, the landslide does not move all at the same time, and back analysis assuming that it does would be incorrect. Hutchinson's[14] (1987) exposition of the causes of large displacement, rapid, movement on pre-existing shears mostly concern landslides that have been reduced to $F < 1$ by external agents. In most cases of reported back analysis, the arguments as to why F should be considered to be equal to 1 are essential.

There is another dimension to this, and that is where you have the geometry of everything and pore water pressures at various times before collapse. There are few cases of this, but being interested in the issue, I have had the opportunity to examine a couple. In general terms, as noted earlier for small and medium-sized landslides, internal deformations seem to begin to be detectable in instrumentation as the Factor of Safety (calibrated against actual collapse) drops from about 1.3 to 1.2, and surface deformations begin to be observable as the Factor of Safety drops further from 1.2 to 1.1. Of course, what is detectable is only detectable if the instrumentation is read and the surface observed, and moreover, the instruments have to be in the right place. I do not advise assuming that the reverse is true, *i.e.* that because the inclinometers had only just begun to kick the Factor of Safety is 'OK'! Why not? Because progressive failure may be happening, or there may be other changes occurring that you are unaware of.

COMPARISON WITH LABORATORY-MEASURED SHEAR STRENGTH

Laboratory tests are done on very small samples, and the precision of measurement should not be confused with accuracy. Soil samples are often mistreated during transport and storage. Moreover, they may not truly represent the materials on site. Some laboratories are better than others. Apparatus effects and errors may sometimes cause misinterpretation of the results.

A reason for uncertainty in the interpretation of laboratory tests is the tendency to do sets of three specimens and then to fit a c'- ϕ' line through them. My practice with ring-shear tests is to do at least five or six points, retesting both for repeatability, and to address 'outliers'. Many effective stress tests are done at higher stresses than in the field for reasons of equipment sensitivity and soil saturation, and small errors in the c'- ϕ' parameters may be compounded into excessive c' when the line is projected into a low effective stress area. Additionally, there is a problem that results from multi-stage testing where there is soil brittleness, which again wrongly emphasises c'. Tests on fills that have incomplete saturation may never entirely remove its effects. Some tests may not clearly define a peak, and the point of failure must therefore be defined arbitrarily. Comparing back analyses (done correctly, but to low precision) with inappropriate laboratory tests (at high precision) leads to dissatisfaction[15] with the former that should instead be pointed at the latter and is a huge mistake.

Analysis of a single slide cannot resolve between the various combinations of c' and ϕ' that yield $F = 1$. Some authors write papers in which they deduce the balance between cohesion and friction by finding the slip circle whose depth most accurately matches that in the field (more cohesion driving it deeper), the position of the critical slip surface is mostly governed by the geology, and clever though the approach may seem, it is an application of geology-free soil mechanics that is likely to lead to some misunderstanding of the real problem.

REMEDIAL WORKS IMPROVEMENTS IN FACTOR OF SAFETY, *F*

In his 1977 description of the back analysis technique, Chandler (see Endnote ii for the reference) wrote that indeed, the method by which the back analysis was undertaken had very little impact on the calculated improvement in stability from the remediation works. Hence, although analyses may produce different F values, the ΔF obtained by analysing the slope before and after remediation is the same. This can only be correct if the landslide is not greatly changed by remedial works, in which case the method of analysis probably does have little effect. I might believe in this in general terms (for comparatively small changes in F), but I would be very cautious if the remedial works greatly changed the overall geometry of the problem.

The assumption of (any) cohesive behaviour also reduces the sensitivity to pore pressure change through drainage as a remedial measure.

I have always put the calculation of internal stresses as a priority in computer programs I have written for the analysis of the stability against sliding on arbitrary slip surfaces, and I make sure that the averages of shear stresses and normal effective stresses are in the outputs, firstly for the whole landslide and secondly for the segment of slip surface in each material. If your software does not do this, then the whole-slide averages can be recovered using a handful of simple tricks. The first is to calculate the average shear stress. If you assume that the soil strength is uniform and purely cohesive, then run the analysis, you will obtain a Factor of Safety for the notional strength you have used. Divide the assumed cohesive strength by the Factor of Safety, and *voila*, you have the average shear stress τ. This is very accurate for a planar slide or even for a slip circle, but a little less so for other shapes. Don't worry; it will be good enough for most practical purposes.

Then, do a run assuming only a frictional strength parameter. Divide the tangent of the angle of shearing resistance that you have used by the Factor of Safety that you obtained, and then take the inverse tangent of the result, and that will give you the *mobilised angle of shearing resistance*. Using that and the average shear strength, you can determine the average normal effective stress σ', because the average shear stress must be the product of the average normal effective stress and the tangent of the mobilised angle of shearing resistance. The average shear stress and average normal effective stress for the slide can then be plotted as a point on the same sort of graph as we use for lab test shear strengths, allowing a comparison between the two.

SUCCESSFUL USE OF BACK ANALYSES

I have employed the back analysis technique with particular success in the determination of the residual strength of a number of landslides in what is thought to be the same general geology and that are slowly moving along pre-existing shear surfaces. In every successful case the ground surface was surveyed and the position of the slip surface fixed by direct observation or instruments such as slip indicators and inclinometers. In many of the cases, a large area of the slip surface followed a single bed, and the remainder of the slip surface was in other clay that was sheared to its residual strength, and therefore only slightly different to the strength on the basal planar shear. Where there were other materials present, their strengths had to be allowed for in some way.

The general approach where there were multiple landslides in the same formation was to treat each analysis as a point on a $\tau - \sigma'$ graph (*i.e.* it is a single 'test specimen') – a best-fit line through the points lies well within estimates of error. While it is recognised that the data and results in individual cases are flawed, the value of the collection is greater than the sum of its parts. (See above for methods of extracting average τ and σ' values.) An important lesson to learn from this is not to attempt to obtain ϕ'_r from any single case: to do so could give a value from 14° at low normal effective stress to nearer 10° at high normal effective stress. Stretching the effective stress range at first seems to show a curved envelope, but as more points are added, some of the curvature melts out of sight in the 'error bars' and known inadequacies of individual data points.

A related problem is to determine the shear strength operative at first rupture of a soil mass in a landslide. However, in general, first rupture (or 'first-time failure') is a complex problem involving progressive failure and multiple geotechnical materials, and any success has to be largely due to good fortune or very careful instrumentation of the slope prior to failure. As a result, there are few good case records.

CONCLUSIONS

Back analysis remains a valuable tool. While it has uses in forensic engineering, the corpus of data from systematic analyses of landslides occurring in a single geological unit remains the best way of identifying and understanding field residual shear strength behaviour. Most of the published data sets available come from the UK, and this technique could and should be adopted more widely. Analyses of single slides cannot resolve the balance of c' and ϕ' and never will. Much dissatisfaction with the method can be resolved by (a) applying it correctly, and (b) comparing the results to lab strengths only when the latter have been correctly executed and interpreted.

Back-analysed strengths have the advantage that they can be used in remedial works analyses with a smaller *partial* Factor of Safety than with laboratory test results, as the volume of soil 'tested' is so much greater, and the analysis encompasses some of the uncertainties between the 'before' and 'after' cases, although uncertainties as to how the groundwater will behave in future still remain. Consequently, a respectable Factor of Safety is still required.[16]

BACK ANALYSES AND OTHERS WITH CONTINUUM METHODS

I wrote, somewhere near the beginning of this chapter, that I would return to the issue of back analyses using continuum methods. Because they predict displacements, these methods do have several more aspects of behaviour from which the governing parameters can be judged – relating to displacement magnitudes, displacement patterns, and displacement rates as well as to collapse (at which they are probably least good, in practice).

With more possibilities also comes the possibility for error, of course, and here are some of the issues that I have seen:

1 Analyses that demonstrate that a pre-existing sheared surface in a slope has no effect on its stability. This is a problem of not being able to replicate the behaviour of a major discontinuity, even with 'slip' elements. The results 'disproved' what we know of ground behaviour, and therefore had to be wrong.
2 Analyses that demonstrate that a topsoil strip causes such failure in the ground due to the release of lateral stresses, so that whatever is built on top has excessively large lateral displacements. This is not our finding in practice.
3 Analyses that used plane strain when the actuality was closer to axisymmetric. I have seen this at least twice: once in the case of a storage reservoir that was nearly circular in plan and to be built from material excavated from inside the reservoir footprint, and once with a tailings facility, again almost circular in plan, but instead of unloading the interior of the bund it was filled. Even axisymmetric would not have been perfect, but would have been better.
4 Rock slopes analysed using parameters derived from the well-known Hoek-Brown formula but ignoring the role of major discontinuities

that were known to exist as faults – but in many cases were slip surfaces of ancient landslides that could – and did – move when the analysis 'proved' that the slope was stable.

5 Finding that the finite-element method could not cope with zero effective stress (at the ground surface) so faking the properties to include a lot of cohesion so FE could cope. (On average, the *combination* of c' and ϕ' used in the FE was the same as in the LE, but obviously the individual parameters were not the same, as the FE properties showed significantly less sensitivity to pore water pressures, and somehow 'proved' that a prediction of failure by LE was never made – and, incidentally, partially disproved the principle of effective stress!).

6 Probably the worst technically, but without real consequences, was a 2D analysis of a slab slide in a limestone quarry. The slab had caught on an unexcavated ridge to one side of its path, which was very obvious in the field. The only way that this could be modelled in the 2D analysis was to assume that the slip surface was strain hardening – but at deformations that made it impossible for the effect to be the result of shear-induced dilation![17] The case did highlight that although, in principle, a back-analysis with a continuum method could be used to match displacements even in a case that had not failed outright, there were many permutations and combinations of inputs, and some would give very peculiar results.

The list could no doubt be extended if I was so minded. However, with great power comes great responsibility, and without the responsibility, the power of continuum methods is greatly abused.

I could, of course, list many of the deficiencies I have seen using limit-equilibrium methods for analysis, and just for balance, here are some examples:

1 Analysis done with newly developed software that contained errors.

2 Analysis of highly inappropriate slip surfaces, while missing the most probable ones.

3 Analysing with slip circles, and thus reporting that thin, weak, layers had little or no effect on the computed Factor of Safety (a slip circle cannot run along a plane).

4. Fitting slip surfaces to a cross section that ignored what was found in the boreholes.

5 Fitting slip surfaces to a cross section that ignored what was found in instrumentation.

6 Using the wrong soil model, for example $\phi = 0$, and thus assuming no effect due to changes in internal water pressures. Of course, the method cannot distinguish between the many possible combinations of c' and ϕ', you need something else to help resolve the issue.

CHECKING THE APPLICATION YOU HAVE USED FOR YOUR ANALYSIS

Checking that you have input the slope geometry and properties correctly is one thing, but checking that the software application you use is correct, and comparable with other software applications is a thankless task. Even with the relatively simple algorithms in limit-equilibrium methods, different computer codes give a range of values. The difficulties are compounded with continuum methods. For example, in the finite-element method (FEM), different codes may use different shapes of elements, different solution algorithms (especially for non-linear cases), and perhaps not even the same soil model (or 'constitutive laws'). Non-linear solutions are therefore extremely difficult to validate, and accepting the results becomes a matter of faith rather than proof.

There is little doubt, however, that continuum methods applied wisely give sensible-seeming strains and deformations, which for some analysts is sufficient proof of their efficacy and value. They are invaluable where a big issue lies with the deformations that might occur in an earthwork even at working factors of safety, thus providing insights as to what readings from monitoring instrumentation are – or are not – something to worry about. Without them, analysts need to employ numerically large Factors of Safety and then to rely on those factors to limit deformations, which is hardly a scientific approach.

Checking the dataset used is simplified with the use in modern software of graphical interfaces and results presentation.

NOTES

1 For example, the plastic analysis of frames. This may be done with a portal frame, but not easily for a multistorey building. The yield line method for slabs is not always used, and as for Hillerborg's Strip Method (Hillerborg, A. (1968) *Proc. Institution of Civil Engineers, 41 (Oct.) 285–311.* – many structural engineers would recoil in horror if asked to use it, as it just seems weird to ordinary practitioners).
2 A good review is given in this paper: Chandler, R. J. (1977) Back analysis techniques for slope stabilization works: A case record. *Géotechnique*, 27, 479–495. Even as far back as 1977, the method was much older than that. But don't believe what he says about ring shear tests and the direction of relative movement!
3 After the Imperial system in which weight and mass used the same units, the UK adopted the SI system of units in which they are different, and although we call it 'the metric system', it is just *one* of the metric systems, of which there are several, and mostly other countries use a different version! EC7 introduced 'mass density' and 'weight density' terminology, and 'weight density' is much better and more meaningful to most people than 'unit weight'.

4 Like the infamous 'Pentium bug' that only arose in very specialised circumstances and probably didn't ever bother anybody.

5 Did anyone really need to say this? Wrong input, wrong answer. Garbage in, garbage out. It has been known for decades. If looking for a reference, then here's one: Duncan, J. M. & Stark, T. D. (1992) Soil strengths from back analysis of slope failures. *Proceedings Specialty Conference Stability and Performance of Slopes and Embankments II, ASCE, Berkeley, CA, 1*, 890–894.

6 In one case, it turned out that the landslide was partly covered with trees, and the air survey was done to the top of the canopy. After ground verification was eventually done, the cross section changed, and the analyses had to be repeated. After the printouts had been 'recycled', I had to do the analyses a third time. Sometimes being paid three times for the same job has its attractions! It certainly beats not being paid at all.

7 These, and other methods of locating slip surfaces, are discussed in this paper: Hutchinson, J. N. (1983) Methods of locating slip surfaces in landslides. *Bulletin of the Association of Engineering Geologists, 20*, 235–252.

8 Probably because this identification is so difficult, Hutchinson's 1983 paper does not dwell on it for long.

9 These include the Selborne experiment, and the Carsington Dam failure (Kennard, M. F. & Bromhead, E. N. (2000) The near-miss that turned into a bullseye. *Forensic Investigation, ASCE*, 102–111). A good record of slip surface shape is given by Cooper *et al.* in 1998: The Selborne cutting stability experiment. *Géotechnique, 48, No. 1, 83–101,* for the Selborne controlled slope failure, where the landslide was sufficiently small for about half of it to be excavated away by means of wide trenches, and a continuous trace of slip surface location was established on the trench wall. Such large-scale intrusive investigations in trenches were also done by Henkel (1956 & 1957) for failures in railway cutting slopes. It is quite possible that the section reported by Gregory in the 1840s (see Bromhead, E. N. (2004) Reflections on C. H. Gregory's new cross landslide of 1841. *Advances in Geotechnical Engineering: The Skempton Conference*, 2, 803–814) was dug out and logged in a similar way over a century before. Who knows?

See also: Henkel, D. J. (1956) Discussion on Watson. *Proceedings of the Institution of Civil Engineers, 5*, 320–323; and Henkel, D. J. (1957) Investigations of two long term failures in London Clay slopes at Wood Green and Northolt. *Proceedings of the 4th International Conference on Soil Mechanics & Foundation Engineering, London,* 2, 315–320.

10 Terzaghi classified slip circle types depending on where the toe broke out as 'base failure', 'toe failure' and 'slope failure', the latter being particularly unhelpful. Hence 'perched'.

11 A reported exception is the Queen's Avenue landslide at Herne Bay, UK (Bromhead, E. (1978) Large landslides in London Clay at Herne Bay, Kent. *Quarterly Journal of Engineering Geology, 11*, 291–304). Investigations by borehole were done over a period of around 30 years. It proved possible to locate to the positions of these boreholes sufficient to enable the three dimensional shape of the slip surface to be established. This happy result was simply a matter of good luck, and the availability of suitable borehole location plans.

12 Hungr, O. (1987) An extension of Bishop's simplified method of slope stability analysis to three dimensions. *Géotechnique*, *37* (1), 113–117. Bishop, of course, referred to something completely different in his paper as 'simplified'.

13 Bromhead, E. N. (2004) Landslide slip surfaces: Their origins, behaviour and geometry. *Keynote paper. Proc 9th ISL Rio Balkema: Amsterdam, 1*, 3–22.
14 Hutchinson, J. N. (1987) Mechanisms producing large displacements in landslides on pre-existing shears. *Memoir Geological Society of China* (9), 175–200.
15 A good example of believing the tests more than the back analyses is here: Deschamps, R. & Yankey, G. (2006) Limitations in the back analysis of strength from failures. *Journal of Geotechnical and Geoenvironmental Engineering ASCE, 132*, 532–536.
16 Not the $F = 1$ recommended by Chandler and Skempton in their 1974 paper 'The design of permanent cutting slopes in stiff fissured clays', *Géotechnique, 24* (4), 457–466 – unless you have instrumented the slope and are out of the way when it fails.
17 Strain hardening was presumably an option in the code developed for some metals and left in when adapted for geotechnical use.

Chapter 6

Water, mainly inside the ground

EFFECTIVE STRESS AND PORE WATER PRESSURES

Engineers have understood that water is the enemy of slope stability for a very long time. Poor old Charles H. Gregory[1] was wrestling with failures of cut slopes on the London to Croydon Railway in a very wet winter of 1841, and when he came to write a paper on his experiences three years later, he mused that the cause of failure might be 'the solvent property or the statical force of water'.

Equally, the engineer for the Southern Railway, faced with movements of the Folkestone Warren landslides, across which the Folkestone to Dover railway runs, had determined that drainage adits dug into the landslide would improve stability. Some had even been dug before the catastrophic movements of December 1915 and therefore before Terzaghi[2] and the latter's *Principle of Effective Stress*, and some as part of the remedial works to bring the line back into operation, but most were dug in the early 1950s. Clearly, there wasn't a big enough effect at first, but major movements of the whole Warren seem to have been inhibited by the drainage works for over a century. In one place the water table was reported to have been drawn down by 60 feet, or nearly 20m.

Gregory's musings aside, there may well be some problems where the effect of water is some dissolution of the mineral content, but his 'statical force' was more to the mark.[3]

However, the whole problem of the effect of water on the strength of soil or rock could only be quantified by application of Terzaghi's *Principle of Effective Stress*. This principle dates back to his classic paper of 1927 and the theory of consolidation. Put rather simply, the central thrust of the principle of effective stress is that all mechanical phenomena in soil mechanics are governed by behaviours that depend either on the *level* of effective stress in the soil, or on the amount of *change* of effective stress in the soil. The theory of consolidation is one of the cases where the *amount* of volume change and therefore surface settlement is a function of the *change* in effective stress, whereas soil strength is generally a function of the *magnitude* of the current effective stress. Soil, being a complicated material with a

 DOI: 10.1201/9781003428169-6

history, may have components of shear strength that do not relate to effective stress but to factors such as past stress levels and mineral cementation, giving a cohesive component to strength as well as the 'frictional' component that is effective stress dependent. The effective stress in a particular direction at a certain point in the soil mass is simply the total stress *minus* the pore water pressure at that point, a simple relationship that probably escaped the thought processes of people before Terzaghi, and even after he had enunciated the principle it took a long time to sink in. Gregory should have been able to do it, but perhaps he thought the matter too complicated instead of how simple it is, or had other matters on his mind like running a railway.

The whole point of the principle of effective stress is that in order to understand how the shear strength of soil is distributed throughout the soil mass or slope it is necessary to understand the magnitude and distribution of the water pressures, and that brings us to the need to understand the seepage of water through soils and rocks.

In the literature, the water pressure in the soil is commonly called the *pore water pressure*, abbreviated to 'pore pressure', the acronym[4] 'pwp' or given the symbol u, although it is much more common to talk in terms of heads of water rather than using pressure units. Very occasionally, other fluids than water are dealt with, for example pressurised gas in the stability of a volcano edifice or a municipal waste landfill, and there is a large and developing literature on partly saturated soils in which pore air as well as pore water pressures are combined. At depth, however, it is likely that the pore space in soils or rocks is saturated. Undersea landslides have saline pore water, and sea water is very slightly denser than fresh water, but the difference is only of minor importance.

To understand the distribution of water pressures in slopes it is probably necessary to consider not simply the steady-state flow of water, but also how unsteady-state flows are generated by changing hydraulic boundary conditions. I think it is also necessary to appreciate that loading on a slope generates water pressures in the loaded soils, and, as these water pressures are out of equilibrium with the hydraulic boundaries, to further understand that they change with time while attempting to reach that equilibrium again. A similar effect is experienced when slopes are unloaded, and here, in contrast to loaded slopes where the water pressures will usually decrease with time, *i.e.* they consolidate, unloaded soils reach equilibrium by swelling, and as a result, water pressures increase with time as the equilibration process marches forward. That, in turn, may be an equilibrium that is never reached, or isn't even equilibrium, as the inputs and outputs of water vary with time. Of course, some water pressure effects like the water thrusts in joints and other free bodies of water inside a slope may be much more readily appreciated than internal pore water pressures.

I have found that the effects on slopes of part submergence are poorly understood, the effects of waves even less so.

Finally, there are methods of communication of water pressures to slope stability analysis software, and long and sometimes bitter experience tells me that this element of slope stability modelling is poorly done. A state of negative effective stress calculated in the course of an analysis often leads to the calculation of negative strength if the programmer has not allowed for that case in the software, and as a result, artesian water pressures should be avoided in stability modelling in the same way that they should be avoided in a completed stabilisation project.

FLOW NET SKETCHING

The graphical technique known as 'flow net sketching' has been a part of geotechnical education at the Higher Education level (*i.e.* degree and technician qualification level) ever since the technique was developed[5] and first applied to the calculation of flow rates through soil, the governing mathematical solutions only being available for comparatively simple cases. It is also possible to extract pore water pressures from a flow net and then to use those water pressures in a slope stability analysis. The flow net is a set of flow lines that are equivalent to the paths followed through the problem by molecules of water, and a set of equipotentials that are isolines of hydraulic potential. Usually, and in the simple cases taught routinely, the two sets are orthogonal, thus forming a net that appears as a system of curvilinear quadrilaterals. Confined seepage cases are bounded by impermeable surfaces that are effectively flow lines, and by submerged surfaces that are themselves equipotentials. Unconfined problems are rather more complex, and add two additional types of boundary condition: a *free surface*, which is a flow line with only atmospheric pressure acting on it (usually the zero datum for pressure measurements) and a *seepage surface* which is not a flow line as water emerges along it, but is still subject only to atmospheric pressure.

Not only is the technique taught, it also makes a good examination question, and so is memorable to all students of geotechnics or soil mechanics! The teaching of flow nets is usually accompanied by the methods for computing flow rates and for computing pore water pressures: the former is rather easier, and therefore mastered better than the latter, and also is less subject to gross errors if the flow net is drawn poorly.

In some taught courses, the sketching exercise is complemented with a sand tank physical model, where dye lines in the flow are seen through a glass-sided tank, or various experiments using analogues for the flow of water, such as electrical potential in a sheet of conducting paper. The former works badly because of capillarity, while the latter is a very accurate analogue for the mathematical solution but isn't a very good match with the reality of seepage, where the permeability often isn't isotropic or uniform.

The fundamental problem is that the flow net method is oversimplified compared to reality. For a start, it is usually based on the idea of soil as uniform, isotropic, and homogeneous, so that the flow is perfectly represented by a Laplacian equation. Stratum boundaries are rarely completely impermeable (although a tenfold or larger difference in the coefficient of hydraulic conductivity – often abbreviated to 'permeability' – represents a huge diminution of the flow rate across a boundary. Moreover, the free and seepage surfaces of unconfined flow nets are gross approximations that completely ignore such effects as capillarity, evaporation or infiltration, effects that are not constant in time and reflect in the field such things as weather, climate, and even the time of day. Some taught courses will enter into such topics as anisotropy and flow through different materials, but field cases are even more complicated than this.

The amount of time and effort devoted to seepage in taught courses is rather small, and textbooks may or may not even cover the topic at all. Even when they do, the examples they present are too few, the same simple cases are often used in several books and, what is more, often contain basic errors.[6] What is even worse is that sometimes, the same basic errors are present in the cases presented in different books! Is this because the authors all misunderstood the principles, often decades apart? Or is it that they rely too much on what they have read in other books?[7]

EMBANKMENTS AND CUTTINGS

Terzaghi's consolidation theory relies on the idea that the application of load to a soil mass generates pore water pressures underneath it, and that the pressures dissipate with time as water escapes from the soil under the out-of-balance hydraulic gradients that are implied by the stress distribution, and thus the load-induced water pressures that result. I think that engineers were already applying the theory of consolidation to settlements, but perhaps not to stability early in the development of geotechnics. The point is that for embankments where the critical failure surface went down into the foundation soils, the water pressures induced by embankment loading are a critical factor in stability and *do* need to be taken into account. As these water pressures dissipate after construction, but the shear stresses remain constant as they are a function of the loading, then the Factor of Safety against sliding *increases* with time after the loading stops. It was decades later that the implications of this were thought through systematically in a paper presented at a fairly obscure conference in Boulder, Colorado, by Bishop and Bjerrum.[8]

Several things flow from the increasing Factor of Safety after construction, and one of them is that it might be possible, by alternating loading periods and then time for consolidation so that the embankment is constructed in several stages, never to reach a condition of instability and therefore to end up with an embankment that is steeper or higher than possible if it were built very quickly without a break for some consolidation to take place. For very large embankments such as dams this may happen accidentally because there is a winter period in which construction is usually halted.

Weak foundation materials are often alluvial in nature and do have a preferential permeability in the horizontal direction due to some sort of layering within the soil. This may mean that water does migrate outwards from underneath the centre of an embankment (where it has little effect on stability) out to underneath the shoulders (where water pressures are damaging) for a short time after construction, and so the precise instant of the end of construction is not necessarily the most critical point in time,[9] but once the worst effects of a lateral migration of pore water are over there is a general increase in stability in this mode of failure.

A brilliant insight of Bishop and Bjerrum was that the converse case, that of the stability of cutting side slopes, behaves in a reverse but analogous way. Excavations *reduce* the water pressures in the remaining soil, and do so to such an extent that the water pressures under the lower parts of a cut slope may even be negative or in suction.[10] This situation also applies to coastal slopes that are eroded by the sea, and to river and lake banks also subject to erosion. The tendency then is for pore pressures to move towards equilibrium by rising. Therefore, the Factor of Safety *reduces* with time. If the engineer is fortunate, the long-term condition is stability, but if the engineer is unlucky, at some stage in this process a failure may occur. The very simple theory accounts for the commonly observed behaviour of cut slopes which is to fail at anything between the moment of excavation up to a hundred or more years after excavation, or to have a 'delayed failure'. Gregory's cutting failed some three years after construction, a result of it being steep, so the suctions were localised in a small zone at the toe, and therefore able to be lost quickly in a rainy winter. Cut slopes built in the Victorian era failing today demonstrate the delayed failure problem.

The time to equilibrate in either case depends both on the scale of the problem and also on the material present. Loading and unloading on gravels and most sands will probably equilibrate as fast as the loading is applied or faster, and so the loading is effectively drained.[11] In clays, it is almost inconceivable the loading can be anywhere near as slow as the drainage rate, so that the change of stress is largely completely undrained. In the case of soils with intermediate characteristics, it is likely that partial drainage will take place during construction. The great advantage of performing stability analyses in terms of effective stresses is that the effect on stability of the generation and dissipation of both positive pore water pressures and suctions can be taken into account, and therefore, the estimates of stability cover all the possible situations.

In principle, it ought to be possible to analyse the very short-term condition using undrained shear strength, and the long-term condition using an effective stress approach to shear strength, thus providing bounds to the problem. Using undrained strengths is seductive, as they are quick and comparatively cheap to determine in the laboratory, and in the slope stability analysis the solution is in many methods not even iterative, so is fast. However, as normally measured and used, the undrained strength is not directional, and the effective stress based strength is, so there is at least that obvious difference. Undrained strengths can, of course, be measured in different directions and used in stability analysis, but this would not be a routine laboratory exercise nor would it be commonplace in stability analyses. My advice is to not expect the results of one short-term analysis using undrained strengths and another using excess pore water pressures and effective stress based strengths to produce the same answer, in either the calculated Factor of Safety or the shape and position of the critical slip surface, and you just have to accept that both approaches are approximations. The shape of the critical slip surface in any case is likely to be the result of the presence of subtle or not so subtle aspects of the geology of the slope and not the critical surface computed for a 'geology-free' uniform, isotropic and therefore totally imaginary soil.

If you have a better stress field calculation than that which is given by the method of slices, it may be possible to compute pore water pressures from the pore pressure parameters A and B, in which A tells you how the pore water pressures respond to shear, and B how they respond to changes in direct stresses. B is always positive, and pore water pressures rise with increasing compressive stresses and fall when compressive stresses are reduced. The B parameter tends to be at or close to 1 in saturated soils. A, on the other hand, may be positive, zero or negative. When A is negative, pore water pressures go down in the soil as it is sheared, and that is a response to the soil increasing in overall volume or *dilating*. The opposite occurs when the soil *contracts*, and there the parameter A is positive. The undrained strength of a soil is a function of its initial effective stress state and how the pore water pressures in it change as the soil is sheared. Note that A in the triaxial test is not the same as A in plane strain or axisymmetric conditions!

UNSTEADY FLOW: LOCALISED LOADING AND UNLOADING

Exactly the same principles that apply in the case of loading and unloading of whole slopes apply to *localised* loading and unloading, and many of the caveats apply too. They are that a load is expected to generate positive pore water pressures underneath it, and unloading by excavation (or an uplift force from a structure) to reduce pore water pressures. Ordinarily, the effects are calculated in limit equilibrium methods just for the slice to which the loading is applied with the pore water pressure change calculated *via* the parameter $\overline{B}$ derived from A and B, but of course, this is a gross approximation. Any form of stress analysis would give the stresses diffusing down into the ground and therefore being a combination of direct and shear stresses, and then the pore water pressure change might realistically be more dispersed and not simply that which the approximate method gives. The pore water pressures resulting would then be distributed in the soil mass. It might be possible to calculate this distribution of stresses and then determine the pore water pressure changes with one of the equations[12] that are available for the job. This would prove to be an advantage if, say, a finite element analysis was being carried out, and the inability to do this represents a serious deficiency in a limit-equilibrium, method-of-slices approach.

The fundamental point is that a localised loading and unloading with the associated respective pore pressure increase or decrease has a behaviour that changes with time. The consequences of the difference between the two is discussed further below. In 1954, Skempton introduced the pore pressure parameters A and B in his *Géotechnique* paper, and in the immediately following paper, Bishop considered how these parameters could be combined in a simplified way, which turned out to be $\overline{B}$. It seemed to him that most reasonable combinations of A and B led to values for the composite parameter that were just less than 1 on loading, and just more on unloading, so that to take $\overline{B}=1$ for saturated soil was always conservative. However, beware of the use of the notation in Bishop's slip circle paper.

UNSTEADY FLOW: CHANGING HYDRAULIC BOUNDARY CONDITIONS

The filling and emptying of a reservoir will make changes to the water pressures and flow conditions in the downstream face, but as the reservoir-full, steady-flow, case has to be managed and is almost certainly the worst-case scenario, then other situations don't normally need to be checked.

Instead, the upstream case is the more difficult one. First of all, think of a rockfill dam with a very impermeable core, something like asphaltic concrete. If the rockfill is very permeable, and the rates of filling or emptying reasonably slow, for example emptying by using the water for agriculture or industry, or releasing it in a controlled fashion, then it is probable that the water level in the upstream shoulder will follow that of the reservoir. If the shoulder is somewhat less permeable, then the water will flow into the shoulder on impounding, and out on emptying. Precisely what water pressures there will be at any instant depends on how large and how quickly the reservoir level changes, and if the process is interrupted or reversed. However, the flow will be relatively uninfluenced by soil compressibility. Then, for a significantly lower permeability soil in the dam shoulder, for example a clay fill or material with a significant clay content, the rate of response will be very slow, and then the pore water pressure change will be more a response to the total stress change than to the change in water pressure, at least initially. Things like rip rap (wave armour) will also drain very quickly, and the total stress change underneath is then also a function of the porosity of that layer.

This leads us to the consideration of partly submerged slopes.

PARTLY SUBMERGED SLOPES, DRAWDOWN AND THE CRITICAL POOL LEVEL

Now, we have to remember that the stability of a slope with an external water load depends on two factors: the support given by the external water, and the effect of the water pressures in the slope. These two effects act against each other to varying degrees. If there are no water pressures in the slope, say for a pond with a liner, then the support effect alone works, and the Factor of Safety increases when the pond is filled, and reduces when it empties.

However, there are few practical applications where there are no water pressures inside a slope which is partly submerged. Commonly, as the external water level creeps up from the toe of the slope, the water pressures inside the slope do more damage to the Factor of Safety than can be offset by the support from the external water. Progressively raising the external water level continues the Factor of Safety reduction until a critical point, termed the 'critical pool level', is reached. The critical pool level marks the lowest Factor of Safety during the process of impounding or emptying. Raising the water level any higher lets the support effect win to some degree. Indeed, in effective stress terms and with a soil that has only frictional characteristics, when fully submerged it has the same notional Factor of Safety as with no water load and no internal pore water pressures.

The critical pool level depends on the distribution of pore water pressures in the slope, and except in the case of a very permeable soil with slow water level changes, the pore water pressures in the slope are different on filling and emptying the reservoir. Obviously, in the case where the internal water levels follow the external water level exactly, the critical pool level is the same whether filling or emptying.

In view of the critical pool effect when filling or emptying a reservoir, it is advisable to do it carefully no matter what the calculations say, in order to be sure that there is enough stability of the upstream slopes of a dam or even the slopes of the reservoir are not to be adversely affected. As the critical pool level is typically lower for emptying than for filling, there is not the consolation that if a slope begins to move one can restore stability by emptying the reservoir again – it just doesn't work like that.

Coastal slopes benefit from the fact that sea water (external) is very slightly denser than the likely internal pore water, although this would not be the same in a rubble mound breakwater, where both internal and external water is likely to be seawater.

Of course, a submerged slope does not really have the same Factor of Safety as one that has no external water, because there are the effects of erosion by currents and waves to consider. A partly submerged slope will have the erosional effects most probably concentrated at the water level, and gradually a beach will be formed.

In his classic paper, Bishop introduced a method of dealing with the two opposing effects in a manner suitable for hand calculation. Perhaps his treatment has led to a lack of understanding of the principles involved. His method confuses many readers, but it is far simpler to do than to understand. To apply it you draw a line that is the continuation of the external water level through the slope. When calculating the weight of any slice that is all or part under that line, you use the submerged unit weight, *i.e.* the unit weight of soil minus the unit weight of water, in the part under the line. Then, imagining that you have a piezometric line, you only count the piezometric head as the part above the line. If the piezometric line is below the projected external water level, you count the head difference as negative. The point of the method is that the simple technique removes from the problem a set of weights and water pressures that exactly counterbalance the external water loading, which can then be ignored. This approach does, however, obscure the fact that there are the opposing effects of external loading and internal pore water pressures.

REPRESENTING PORE PRESSURES IN STABILITY ANALYSIS

Typical software applications for slope stability usually have multiple methods for representing water pressures. The simplest method of all is the *piezometric line* – it would be a piezometric *surface* in three-dimensional analysis. This line (or surface) represents the levels that would be reached in a set of piezometers of the open-tube type. The program computes the pore pressure at any point underneath the line as the static head based on the vertical separation between the elevations of the piezometric line immediately above the point in question and of the point. It can be a very economical way to specify the water pressure distribution, requiring only a small number of coordinates to be specified. The downside of the method lies in the fact that it presumes a hydrostatic distribution of water pressures beneath the line, with no allowance for the effects of upflow or downflow, which by rights should respectively demand an artesian or subartesian[13] pressure distribution in the former case, and subhydrostatic in the latter. It is most useful where the analysis uses a single slip surface or a few surfaces close together, and is therefore least useful or accurate when many slip surfaces with radically different depths are used in the computer model.

It is not obvious where the origins of the pore pressure ratio r_u are, but it is a simple concept that has been responsible for a great deal of simply wrong stability analyses. At a point in the slope, this pore pressure ratio is defined as the ratio of pore water pressure to vertical total stress, the latter computed simply as the depth below ground surface multiplied by the soil unit weight. If necessary, this approximate vertical stress can be computed by summing the effects through multiple layers of soil. The ratio could also be computed as an average along a surface such as a slip surface, or over an area or throughout a volume. I long suspected that the origins of the method went back to Little and Price, who needed a simple and very compact method of both specifying and storing pore water pressure information for their computer program and processing it with a minimum of code – within the program the pore water pressures are recovered from the pore water pressure parameter and the vertical stress. However, the concept is older than their paper by a few years. A simple method of the r_u kind is also required to manage water pressures when using stability charts.

The use of a pore pressure parameter defines the variation of water pressure with depth and across a stratum in a way that is non-intuitive. It also tends to hide both high and low pore water pressure zones and thus may affect where the critical slip surface is predicted. It is also found that the distribution of pore water pressures is different to the subdivision of the slope section into different materials, so a further subdivision is required. While in principle a very fine zonation of the cross section allows the pore pressure distribution to be represented through the use of different pore pressure ratios in every zone, the distribution is also discontinuous at every boundary. It is also difficult to make small changes to the geometry, say by placing berms on the surface, without changes appearing in the pore water pressure distribution, even if the pore pressure ratios are adjusted.

There is a further complication in that Bishop used the symbol $\overline{B}$ (B-bar) instead of r_u in his classic paper. $\overline{B}$ is the increment of pore pressure as a proportion of any *increment* of vertical total stress when loads are added (or subtracted), whereas r_u is the ratio of the *accumulated* pore water pressures and vertical total stresses. Bishop's example (which possibly controlled his thoughts on the matter) included an earth embankment where the increment of each variable starting from zero was the same as the accumulated value. A few years later in his stability chart paper with Morgenstern, Bishop adopted the more conventional nomenclature used from then to today.

Personally, I dislike the use of r_u, and although a value of zero is useful to signify the absence of water pressures, such a case can usually be represented by other means.

Something I also dislike is to define a piezometric line for each soil zone. This implies a discontinuous water pressure across each soil boundary, and a hydrostatic variation of water pressures with depth in each zone.

A leading commercial software application for slope stability analysis is part of a suite of geotechnical modelling applications, one of which is a finite element–based seepage analysis, the results of which can be interfaced with the slope stability calculations by means of a grid of points at which the pore pressures are specified. It is a method that has a lot to commend it, but my own favourite is to assign pore water pressures to the nodes that define each soil boundary, and to perform interpolation in the program. It still takes some care to get the distributions right, but the checking is assisted if the program can contour the pressures or pressure heads to show what the interpolation implies. Discontinuous pore water pressures can be specified by placing two coincident lines with a zero-thickness soil between them. When using this method in a big zone like a dam core with many piezometers, the zone can be divided into subzones by lines that string together the piezometer positions.

I see that I was using this method in the early 1980s,[14] and I cannot now remember if I invented it or simply took it from somewhere else. However, it is part and parcel of an approach that I recommend, which is to separate the water pressures in an analysis from their causes such as external water levels, loading and unloading, *etc*. This makes the analyst think about how the water pressures generate and dissipate. It also avoids a struggle about how to represent certain flow conditions, and the accidental modification of pore water pressures under an embankment added to a soil model where the pore pressures are computed *via* r_u.

Where the complete pore water pressure field is not available, I tend to use the piezometric line, especially in the back analysis of landslides in which the position of the slip surface is known, and so are the pore water pressures along it, however approximately.

THE PROBLEM OF PONDS

Ponds on landslides pose a problem in that the water has weight, but without careful observation and perhaps testing it is not clear whether or not the pond has an adverse effect on stability. If we consider a pond (or a swimming pool, say) with an impermeable liner, then the weight of contained water acts like any other load: adverse if upslope of the neutral point, but beneficial if downslope of the position of that point. It is not therefore always helpful to drain ponds, especially if they are low down on the slope and have a water level that is independent of the groundwater.

In the case of ponds without an obvious surface water inflow, it is likely that the pond level simply reflects the water pressure heads in its vicinity, and therefore there would be little benefit in draining it. A spring-fed pond may even have a lower water level than in the soil at depth, especially if it overflows and discharges into a stream or channel. However, a pond that is kept full by surface water flows (or is fed from a mains water supply) is probably a nett contributor to groundwater levels in its vicinity.

Ponds often form on the surface of landslides in low spots if the ground surface is deformed.[15] These are usually fed by rainfall, broken drains and other pipes, and so in many cases it is beneficial to drain the water away from the surface of a fresh landslide no matter where the pond is located.

Incidentally, does the weight of a pond contribute to the calculation of pore water pressures with r_u? If it does, then r_u at the base of a pond is 1, and rapidly decreases with depth. If it does not, then the r_u at the base of the pond is infinite. It is probably worth abandoning r_u in such cases. I certainly do so.

CONTINUUM METHODS SUCH AS THE FINITE-ELEMENT METHOD AND SEEPAGE FORCES

Pore water pressures in soil can be dealt with by considering *seepage forces*, or drag against the flow of water, as an internal body force. Such a treatment is an alternative to using pore water pressures but having the same result. In methods like the limit-equilibrium method of slices it would be an unconventional approach to use seepage forces, but methods such as the finite-element method already contain the computer code toolkit to evaluate the hydraulic gradients that give rise to seepage forces and then to integrate them over the area (or volume) of elements. One should not forget that the unit weight of water must be subtracted from unit weights within the seepage domain, but only once, as changes in the hydraulic gradients are used as changed body forces.

It is a mistake to believe that seepage forces and pore water pressures are in any way different: if you are conducting an analysis that considers, say, the pore water pressures on the sides and base of a slice in a limit equilibrium analysis, the pressure differences around the water body in that slice lead to the calculation of a nett force that is the same as if you convert the hydraulic gradients into seepage forces and use those instead. You can have one or the other, not both.[16]

Pore water pressures are sometimes referred to as 'neutral pressures'. In my experience, they are anything but neutral, and are on the side of the enemy!

SUCTION

Suction, or negative pore water pressure, occurs in soils for a number of reasons and through a number of mechanisms. One of the mechanisms is the unloading that occurs when a cutting is excavated or a coastal slope eroded into shape. You should note that with the passage of time, such suctions may be gradually replaced by zero or positive pressures as the slope moves towards an equilibrium set of pore water pressures. I would like to say that the seepage in the slope equilibrates to a steady seepage condition, but steady seepage is unlikely ever to be reached, as slopes are subject to seasonal water pressure conditions, depending on whether the face has nett infiltration or evapo-transpiration.

Another mechanism that gives rise to suctions in clay fills is the effect of remoulding during compaction. Clay fill embankment slopes therefore have rather different time-dependent stability behaviour to the embankment and its foundation acting together. The latter mechanism generally gets more stable with time, as discussed previously, but the embankment slope may lose its suctions in the same way as a cutting slope does, and thus be subject to delayed failures. Steep embankment slopes can throw off a lot of precipitation due to their steepness, whereas flat slopes may not be so lucky, and more infiltration occurs. However, as steep slopes are intrinsically less stable and flat slopes more so, there is an intermediate range of slopes for which the stability is least. Vegetation on embankment slopes is usually helpful in maintaining suction, but fallen trees and the 'wrong kind of leaves'[17] encourages clearance, with a corresponding loss of suction.

Suction is also found where the ground has been heated, *e.g.* underneath brick kilns, or frozen, *e.g.* underneath industrial cold stores. Neither is necessarily likely to form a future slope, but the process also has a time-dependent recovery of pore water pressures, volume changes, and a loss of strength.

Long, periods of dry weather, particularly where they occur frequently, as in countries with a dry climate, can also induce or maintain suctions in soil slopes.

My personal view of soil suctions is that they are an interesting phenomenon that often explains why some slopes are stable, and indeed, why delayed failures take so long. Suctions do dissipate, and to rely on them seems to me to be, in effect, relying on how hard the ground can suck and how long it can hold its breath. Occasionally, one may need to rely on this tendency when making temporary excavations, for instance, but I would rather not. It is, for example, difficult to calculate just how long an oversteep excavation in a stiff clay will stay open, and the answer may not be entirely influenced by pore water pressures, but also by the joints and fissures or other discontinuities in the soil mass. This uncertainty is one of the reasons we do not enter trial pits nowadays.

THE WATER BALANCE

I find it useful when dealing with a large natural landslide or a big earthfill like an old colliery waste tip to consider the water balance, or what the water inputs and outputs are. Obviously, an excess of inputs would mean that the water body was growing, and the reverse that it was shrinking, but I have never had the necessary data to do this in more than a very approximate way. Generally, taken over a time interval of years, the inputs and outputs must balance, but clearly, over a short period they do not, and the water body sometimes grows and sometimes shrinks. In the UK with (allegedly!) a different rainfall pattern in winter and summer, there is a seasonal effect. The idea of a water balance is useful, especially if approximately quantified, because it shows whether the relative impact of human factors is significant or not, and has the other benefit, once all inputs and outputs have been identified, of leading your mind to what factors are more controllable and therefore to ideas of actions that might improve stability.

The exercise of identifying water sources such as perennial ponds and springs shows where the water table lies, at least locally, and identifies the matching sinks. If springs or sinks are found, they show where the water table is higher or lower than ground level. Occasionally, the names of properties give some insight, as in 'Spring Cottage', 'Winterbourne' (a *bourne* is a seasonal spring) or 'East Dene' (a *dene* is a sink), but it is always worth checking, as the naming of properties often veers towards the fanciful ('Fairy Dell', anyone?).[18]

As for seasonal groundwater conditions, vegetation is sometimes helpful, as some plants need marshy conditions in at least part of their growing cycle, but when you are up to the tops of your Wellington boots in mud in the middle of the summer, you may not need such clues.

Plants that I find useful as indicators are the biennial teasel (*Dipsaceae*), reeds (*Juncus*) and horsetails (*Equisetum*).

UNDERWATER LANDSLIDES

Underwater landslides have the potential to cause damage and could cause, or be caused by, tsunamis. Several case histories are available, including the progressive severance of transatlantic undersea telegraph cables in 1929 in the Grand Banks off Newfoundland, the discovery of the huge Storegga slide off the coast of Norway, and some sea bed mapping in pursuit of oil exploration. At the time I write, there isn't much an engineer can do about them, and in practical terms the publications about them tend to fall into the category of *'I've found a bigger landslide than you'* rather than *'This is how we fixed it'*, probably because it didn't need fixing, and even if it did, there was no practical way of doing it.

Sea bed landslides have occurred throughout geological history, and sometimes, their effects are found in sediments of considerable age. For example, slip surfaces from submarine landslides that had occurred in the Carboniferous were found, and caused difficulties[19] during tunnelling as part of the Carsington scheme. Other ancient landslide slip surfaces may be preserved in the sedimentary record, and be described by a geologist as a fault, listric surface, or some other term[20] where the soil or rock fabric shows evidence of being sheared.

A train of high-amplitude, long-wavelength, oceanic wave may be caused by an intense storm, and running up an underwater slope that may represent an extreme case of loading at the head and unloading at the toe, and in combination with cyclic loading-induced pore water pressure change that causes failure in the submerged deposits resulting in a landslide. To come full circle, a submarine landslide can cause large waves, as can phenomena such as rapid fault displacement. Landslides from land into lakes, reservoirs or landlocked bays can cause huge waves, and there may be a case for assuming that a landslide into a relatively unconfined area of the sea might cause a *train* of large waves.[21]

GEOFABRIC FILTERS AND SEPARATORS

Prior to the invention of geofabrics, drains could only be filtered by being constructed with a zone of soil which has an intermediate grading between the soil and the drain itself. Multiple layers may have been called for. The geofabric filter provides a 'one-stop shop' to separate the soil and the drainage medium. This permits drains such as trench drains to be provided with a filter, whereas in previous practice they were not, for reasons that included the lack of space. Graded drains with multiple zones are used in dams, where there definitely is room.

Geofabric filters may also be provided for perforated pipes and other drains, but they always pose a problem of maintainability.

On balance, however, I think that I prefer drains to have geofabric filters than that they should not. Time will tell if that opinion is right.

DESIGN CODES

Design codes often say little about groundwater, although EC7 exhorts you to consider the worst probable water pressures in the ground when preparing a design. If this is taken to mean that you need to consider what happens if the drains fail, then one will find that extremely flat slopes are required in any design! Otherwise, it is a sensible suggestion. In particular, it should caution you against being too eager to use a design based on suction! (Unless, of course, you can guarantee that it will never rain on your site for long enough to destroy those otherwise useful negative pore water pressures.)

In order that the design water pressures are the worst experienced in slopes where drains are provided, it is probably best to monitor the water pressures in-situ and also to over-provide drains, as in may cases they cannot be replaced. In the cases where they can, an inspection regime and timely maintenance are essential.

Drainage as a stabilisation measure is one of the topics discussed in Chapter 9.

SURFACE WATER

Finally, I think that it is worth remembering that the movement of surface water causes erosion, something that merits the coverage of the following chapter. This is most obvious in the case of the erosion of the toe of a slope, for example, a coastal cliff. In the case of a 'soft cliff' or the erodible slopes around the coast of SE England, the toe erosion has probably created a wave cut platform associated with present sea level, but there are places where the true toe of the slope is well below present sea level and the slope has been drowned by rising sea levels following the last glacial retreat. In such cases, the roughly 6ka stasis in sea level up to the present may mean that contemporary erosion is cutting into the middle of a landslide system, undermining what is possibly the less stable upper part by separating it from the more stable lower part. A similar effect may be experienced in a reservoir that drowns the lower part of the valley slopes.

Water running down stream valleys may pick up soil and rocks which interfere with the use of transport or water supply infrastructure, or block culverts, carry away structures, or cause other disruption, the effects of which may variously be attributed to floods rather than landslides. Flood waters in desiccated clay terrain may cause dispersion at a significant rate, and such terrain develops pipes. The dispersivity of some clays may be simply a function of wetting up or may be due to the mineral properties. Regardless of which is the cause, the surface water rapidly becomes a flow of mud that for all its mobility on a slope becomes capable of blocking drainage systems at all gradients. The resulting 'badlands' topography is especially well exhibited in southern Italy, where it is known as 'calanchi'.

NOTES

1. He wasn't that old at the time. In later years he served as president of the Institution of Civil Engineers and has his portrait on the first floor of the Institution building in Westminster. As the building often serves as a backdrop surrogate for a palace in TV dramas, I often see him smiling down sagely on the actors! His paper is: Gregory, C. H. (1844) On railway cuttings and embankments: With an account of some slips in the London Clay, on the line of the London and Croydon railway. *Minutes Proceedings of the Institution of Civil Engineers*, 3, 135–145. The paper was one that was reprinted in *'A Century of Soil Mechanics'* by the ICE, but the ensuing discussion in the original journal is also informative.
2. Terzaghi did visit the site during his visit to the UK immediately before the Second World War, when investigations were taking place for a phase of remedial works that wartime delayed.
3. As I discussed in my paper: Bromhead, E. N. (2004) Reflections on C. H. Gregory's new cross landslide of 1841. In R. J. Jardine, D. M. Potts, & K. G. Higgins

(Editors) *Advances in Geotechnical Engineering: The Skempton Conference*, vol. 2. Thomas Telford, 803–814.

4 Some pedants will insist that this is not an acronym, as it cannot be pronounced, although according to the Oxford English Dictionary they are wrong.

5 Brits will be pleased to note that this was developed by L. F. Richardson in 1908. Americans can continue to believe it was by one of the Casagrandes. (Richardson, L. F. (1908) The lines of flow of water in saturated soils. *Proceedings of the Royal Dublin Society*, *11*, 295–316). King, F. H. (1899) in his *Principles and Conditions of the Movements of Groundwater*, 19th Annual Report of the US Geol. Survey, Part II, Washington, could draw a flow net for flow in a horizontal plane, but not in a vertical one.

6 I wrote a paper about this: Bromhead, E. N. (2007) Flow nets and text books. *Proceedings ICE Geotechnical Engineering*, *160* (4), 203–207. I would never have discovered how widespread the problem was but for a student pressing me to show him the best examples in my preferred book. To my astonishment, I needed to go through my whole collection of textbooks, those belonging to colleagues and a library before I found a book devoid of errors! That book was Cedergren, H. H. (2009) *Seepage, Drainage and Flow Nets*. Third Edition. Publisher Wiley – any one of the three editions will truly open your eyes. Wu's flow nets with errors were in the example exam questions where the student was expected to point out the error. (Wu, T. H. (1969) *Soil Mechanics*. Fourth Edition. Allyn & Bacon). Some textbooks do not cover seepage at all, and many miss out slope stability.

7 I couldn't help but be reminded of the character Adso in Umberto Eco's *The Name of the Rose*, when he finds out that the purpose of the library is not to show the truth, but sometimes to conceal it. But he also remarked that books speak of other books, so that one can reconstruct the contents of a lost volume. Certainly, this is true of the coverage of flow nets in the average textbook!

8 Bishop, A. W. & Bjerrum, L. (1960) The relevance of the triaxial test to the solution of stability problems. *Proceedings of the ASCE Research Conference on the Shear Strength of Cohesive Soils, Boulder, CO*, 437–501.

9 During a trial embankment test in the Thames Estuary, the monitoring team concluded that they had not induced failure, so they packed up and went to the pub. The embankment failed while they were away.

10 In the Selborne controlled slope failure experiment, notwithstanding the injection of considerable volumes of water into the slope, the suctions at its toe were never fully eliminated. In the coastal slopes of the Isle of Sheppey in the London Clay Formation, the failures occur with pore water pressures in large parts of the slope still in a state of suction.

11 Drained as in the triaxial test, not whether or not the site is drained as a stabilisation measure.

12 There is, of course, the Skempton pore pressure equation: (Skempton, A. W. (1954) The pore-pressure coefficients *A* and *B*. *Géotechnique*, *4*, 143–147). The coefficients were developed for use in the triaxial test, and the parameters do not translate easily from the test to the plane strain conditions of a section through a slope. For example, *A* for an elastic soil is 1/3 in the triaxial test but ½ in plane strain. Other formulae might be used, and of course, the results will depend on what is done.

13 *Artesian* means that the water head rises above the ground surface; *subartesian* means that the water pressure head rises to above the piezometric line (if used in a computer model) or groundwater table in the field.
14 In particular, I used it to analyse the stability of the Carsington Dam using both Bishop's Routine Method and the Morgenstern-Price Method. I found that the factor of safety indicated probable failure. The following year, as the dam earthworks approached completion, the dam failed, the warnings given having been completely and deliberately ignored. See for example, Kennard, M. F. & Bromhead, E. N. (2000) The near miss that was turned into a bullseye. *Forensic Investigation, ASCE*, 102–111.
15 The infamous landslide at Hawkley, described in Gilbert White's *A Natural History of Selborne*, changed the shape of the landslide surface so the pond was deep where it had been shallow and *vice versa*. It is a hint that the landslide already existed and was just reactivated, rather than being a fresh failure. White himself was ill in bed at the time and sent his nephew to report on the slide. He later took his nephew's account as his own in the book. Shame on him, and him being a clergyman as well!
16 If you believe that they are different, and therefore need to be added or combined in any other way, then you are making a big mistake – a mistake that De Mello and colleagues made in their paper to the Skempton Memorial Conference. It just goes to show that no matter how august you are, it is possible to make a fool of yourself.
17 A spokesman for the UK Rail Network once used this phrase in an interview, to great derision. What he meant was that sometimes the leaves reduce friction between steel rails and steel wheels. There is no 'right kind' of leaves.
18 Typically, you won't find small, magical, winged humanoids. You also are unlikely to find a pot of gold in the garden of 'Rainbow's End'. But, you might find landslides at Oakfield Road – I have, several times. Oaks apparently like to grow in clay soils, and clays – well, you might get the gist.
19 As reported by Norbury in his Glossop Lecture: Norbury, D. R. (2017) Standards and quality in ground investigation: Squaring the circle. *Seventeenth Glossop Lecture. QJEGH*, *50*, 212–230.
20 If you don't understand a technical term used by a geologist, then don't be inhibited from asking for an explanation. After all, they don't understand all the terms you use. Sometimes you will remain confused, and so will they. Even if you understand a term, it may be used in a different way by a geologist from a different country. This assumes that I am talking to a civil engineer. Geologists should remember that many civil engineers only ever had a dozen lectures on geology, and may well have been asleep at the time.
21 Radar stations in the Gulf of Mexico anchored in the coastal slopes were thought to have been towed under by landslides. The waves from the Vaiont reservoir caused by the landslide were extreme, as were the waves in Lituya Bay from another landslide. It is highly probable that landslides from the flanks of Hawaii or the Canary Islands or elsewhere could cause huge waves at source, but how they might propagate is unknown.

Chapter 7

Surface water in motion

Scour, transport, and deposition

INTRODUCTION

While concentrating on the effects of water pressures *inside* the soil or rock mass and their effect on shear strength and stability, the previous chapter has begun to deal with some of the issues of surface water, for example by considering ponds situated on a slope and thereby contributing their weight to the overall stability equation. Sometimes, as noted in the previous chapter, with ponds on the upper part of a slope, their weight contributes to destabilising the slope, but the weight of ponds low down on a slope may help stabilise it. However, if the lower-slope pond causes inflow to the slope, the effect is positive or negative depending on the magnitude of each effect. Upper slope ponds are always injurious. There is also that conundrum about the inundation of the lower part of the slope and whether it improves or reduces stability. It is worth reminding ourselves that the water load on the lower part of the slope always improves stability, but the creation of water pressures *inside* the slope material must reduce stability, so that these two effects fight each other, and whichever one wins is a function of the shape of the slip surface and the pre-existing water pressures inside the slope before the inundation occurs. A further consideration about the weight of a water body on a slope is when the application of the weight causes undrained pore water pressures, which also impact stability negatively.

However, in the previous chapter, all the water bodies were more or less static, and matters change significantly when the water is moving, which gives it the power to *erode* the soil or rock surface, to *transport* the resulting debris, and then to *deposit* it somewhere where it will more likely than not prove to be an inconvenience.

During university courses in civil engineering, slope stability is taught as part of soil mechanics or the more generally termed geotechnical engineering, but various forms of erosion are rarely touched on, except, if the students are lucky, during a field visit or as part of a student project. In the following sections, the problem of surface water erosion is described in the form of the different scenarios in which surface water erosion, transport or deposition takes place.

DOI: 10.1201/9781003428169-7

COASTAL AND RIVERINE SLOPES

A very obvious problem of erosion occurs in the case of a coastal slope, where the sea is the principal agent, acting through both waves and currents, chiefly at higher states of the tide. The sea acts at the lower part of a coastal slope in response to tide, and it may be that for a large part of a tidal cycle there is no direct attack. At high tide, a slope may be subject to currents and waves, with the effect of each modified with both different weather patterns and the lunar cycle affecting the tidal range. In contrast, slopes down to rivers may be subject to a range of discharge rates in the river, but flow is usually in the same direction, and the effect of waves is minimal.

It is appropriate here to remind you of the wave cut platforms associated with periods of sea level or reservoir level stasis (mentioned at the end of the previous chapter), which may have been cut into the body of an ancient landslide system. They can be as problematic as present day erosion at the toe of a contemporary slope.

The transport of material along a coastline is usually referred to as *longshore drift*, and it is only a stability problem if it also causes erosion *en route*. Of course, the material has to be sourced from somewhere, and if that source is cut off by coastal protection works, then the longshore drift may well turn from a fairly neutral process to one that is damaging. Attempts to inhibit the movement of materials often take the form of fences or *groynes* constructed at right angles to the shore, which capture the moving sediment. Groynes have traditionally been constructed of various types of timber, steel sheet piles or concrete, and, in some cases using rock blocks, with a requirement for maintenance that varies depending on how aggressive the sea and its fauna are in a particular location. Harbour arms act as super-groynes in a coastal context where there is longshore drift.

For obvious reasons, analogues to groynes cannot be used in rivers, and river banks can only be protected by some form of wall, with stone-packed gabions being a particular favourite since they do not significantly block water flows from the slope into the river. Gabions are much less frequently used as seawalls, with masonry or concrete being preferred, and steel sheet piles only used sparingly, as alternate wetting and drying of the steel leads to fast corrosion. Masonry and concrete sea walls are often constructed with a profile to deflect the wave energy, but the returning wave is erosive, so that sea walls of this kind usually require an apron with possibly a toe (in which case sheet piles come into their own) to protect the whole wall from being undermined. Where the sea wins over man, it may be necessary to further protect the wall and its toe with rock or concrete armour blocks to dissipate the wave energy.

Substantial sea defences may trap beachgoers as the tide rises, and suitable means of escape from the beach must be provided. Not a stability issue *per se*, but nevertheless an important design element for public safety.

RUBBLE MOUND AND OTHER BREAKWATERS

Since the whole point of a rubble mound breakwater or indeed of any breakwater is to attenuate wave heights inside the protected harbour, it is inevitable that the peak of a wave on the weather side is likely to be substantially higher than the water level inside the harbour, thus giving an overall hydraulic gradient from weather side to lee side. The hydraulic gradient is at its most steep immediately underneath any wave wall structure built at the top of a rubble mound breakwater, and it is a fundamental design or construction error for the breakwater to be permeable at this location, as the water flows may erode the lee side of the breakwater as the water emerges. The opposite problem of flow from lee side to weather side is not a problem even if waves have a trough elevation lower than the water level in the harbour, because firstly the levels are closer to each other since there is no analogue to wave run-up, and secondly, there is usually some intermediate layer between the armourstone and the body of the breakwater to act as a filter as each wave retreats.

Should the wave wall be insufficiently high so that waves break over it, the resulting flows over the lee side of the breakwater may cause erosion, since almost inevitably it will be provided with less in the way of wave armour than on the weather side. The erosion problem is compounded if there is a combination of both through flow and overflow.

The conventional solution to wave attack along the weather side of a breakwater lies in the provision of armour in the form of randomly placed natural stone or concrete blocks, or of special shapes of concrete units, each with its own proprietary name. A common feature of these concrete armour units is the complexity of the formwork in which they are cast, leading to a slow production rate in the casting yard, which may need to produce hundreds or even thousands of units before they can be deployed effectively, and the heavier units need ever larger cranes to lift and place them.

Offshore breakwaters are lengths of rubble mound placed parallel to a shoreline, but with short gaps between them, intended to reduce the wave energy actually acting on the shore. Ideally, stable embayments then form inshore of each opening, in some cases with artificial tombolos connecting the breakwaters to the shore. Some beach nourishment (importing sand or gravel as appropriate to supplement the natural beach) inshore of the breakwaters is helpful in many cases.

CUTTINGS AND EMBANKMENTS

The amount of precipitation falling on an infrastructure cut or fill slope or arising from snow melt is not usually enough to cause significant erosion, except when the slope is newly formed or topsoiled, and before vegetation has become established. Protection in such cases is easily provided with a thin mat of a biodegradable material such as jute. More complicated slope profiles with flat areas such as berms, and slopes of considerable area such as reprofiled coastal slopes or the faces of large dams may need specific drainage measures to capture and control runoff.

Generally, however, small earthworks require only protection from water that has been accumulated elsewhere and then runs over the surface in a concentrated flow. The capture and transport of surface water is a separate matter to the capture of seepage from inside the slope, but it is difficult to create separate systems, and the result is commonly a system such as chevron or herringbone patterns of surface water collector drains feeding into longitudinal drains that perform the deeper drainage function as well as transporting, and ultimately discharging, the collected water.

A question to which I will return in Chapter 9 in connection with the disposal of the collected water is 'where to?'. The question is easily answered for most coastal slopes, where it is 'Into the sea, of course!', but in other scenarios it is by no means so obvious. Indeed, even in a coastal setting if the water has been collected from a village where all the houses are on septic tank drainage, the water may well be contaminated with some bugs, notably *e. Coli*, and therefore requiring treatment before discharge. Chemical and other contaminants (for example from mines and quarries or other industry) may also need pre-treatment before release.

Runoff from an inland slope is therefore more complex for a variety of issues. Consider, for example, a cutting on a long decline leading to a railway tunnel. It is then possible that the collected water has to be channelled through the tunnel for discharge on the other side – impossible if the far end of the tunnel meets another cutting with the line rising! In such cases there is little alternative to a pumped discharge solution, with all the operating energy and maintenance costs to factor in.

WHAT DAMAGE CAN A SURFACE WATER FLOW OVER A SLOPE DO, AND HOW?

When water passes over the surface of a slope it applies a shear to the surface related to the square of its velocity, and at the same time, the pore water pressures for a small distance inwards from the slope face rise to extreme states that are usually enough to dislodge some particles. In the case of rock slopes, the size and shape of particles is determined by the nature of the discontinuities present in the rock mass, their spacing, persistence, and orientation. Clearly, massive rocks are less susceptible to being eroded than are weak rocks or anything tectonically sheared already into gravel-sized or even smaller pieces. Additionally, the transportation power of the water flow increases dramatically with the discharge, and that in turn is a function of how the overland flow is concentrated by the shape of the source landform.

Some soils are dispersive, *i.e.* their particles readily part company with each other, and the process of dispersion is accentuated with unsaturated soils because wetting fronts advancing into the slope both increase pore air pressures and dissipate surface tension effects. The nett effect of the entrainment of soil particles may be to create a wet flow or to undermine and release larger particles such as boulders or the root balls of vegetation. Flood waters in desiccated clay terrain may cause dispersion at a significant rate, and such terrain develops pipes. The dispersivity of some clays may be simply a function of wetting up, or it may be due to the mineral properties of the soil. Regardless of which is the cause, the surface water rapidly becomes a flow of mud or debris that, for all its mobility on a slope, becomes capable of blocking drainage systems at all gradients.

Water running down stream valleys or simply over slopes may pick up soil and rocks, and carry the detritus to places where they interfere with the use of transport or water supply infrastructure, or block culverts, carry away structures, or cause other disruption, the effects of which may variously be attributed in reports to floods rather than landslides. The moving mass of soil and water is often described as a *debris flow*. Alternatively, it may be the cavity left after the removal of material which is the problem, usually (but not always) in the case of water-retaining structures where their integrity is breached. The slopes are, however, unstable in such situations, and the job of the relevant engineer is ideally to prevent the occurrence of such damaging phenomena, or to repair the damage caused after the event. On the grounds that prevention is better than cure, the former is to be preferred to the latter.

RAILWAYS – MAINLY (BUT NOT ALWAYS) A PROBLEM WITH CUTTINGS

Sloping earthworks on a railway network are often of considerable age: in some cases approaching 200 years old. Should a cut slope fail it will deposit material onto the tracks, and in the alternative case where a filled slope fails it may undermine the tracks. In both cases, a train approaching at a moderate to high speed will probably be derailed, and in the case of a cut slope that may mean impacting a train coming in the opposite direction; hitting the portal of a tunnel; or, if the cutting leads in the direction of travel to a bridge, the bridge parapet. In the case of a filled slope failure, a derailed train will probably plummet down the side of the embankment, possibly impacting trees on the way, with the possibility of inflicting further damage to whatever is situated at the toe of the slope, and to the train itself, its passengers, and crew.

Several factors contribute to the seriousness of these events, which are thankfully rare, although frequent enough to feature in the reports of the Rail Accident Investigation Branch (RAIB)[1] in the UK and to feature on the organisation's website. Incidentally, there is no point in looking for the keyword 'landslide', as in railway parlance they are 'landslips', a term that was the common one until comparatively recently. Related investigations in the US are reported by the NTSB. One of the factors is that the train driver rarely has a big enough sight distance needed to stop the train at the speeds at which they travel and even less at night or in inclement weather. Another factor is that there are never 'crash barriers' (as there often are on the road network) to keep the train from plummeting over an embankment or hitting another post-accident obstruction. The amount of material that needs to be deposited on the rails to cause a derailment is comparatively small, and if a train cannot stop in time, it also cannot swerve. In fact, the driver of a train is critically dependent on the signals that are set from a remote signal box, where the signalman may not even be aware that there is an obstruction on the track. One might think that modern trains with the driver positioned at the front with a clear view of what is ahead might be rather better than the case of steam trains, where the driver's view ahead was restricted greatly. Everyone of my generation who ever had a train set as a child would probably have wondered how on earth the driver of a steam train was expected to look through a small porthole and see where the train was going, and then rationalised it that the driver didn't need to as the train followed the tracks. One watches the dramatisation of *The Railway Children* and breathes a sigh of relief that Bobbie Waterbury (the actress Jenny Agutter) was indeed lucky that the driver was actually looking, and the train wasn't going very fast![2]

It does not take a substantial landslide such as the failure of the slopes near New Cross (or, indeed, the many other failures that were a subject of great interest to Skempton and his acolyte Henkel, with far too many cases to mention individually) to obstruct the track; indeed, at the time I write this there is no such event recorded in the reports of the RAIB, although the reports online only start *circa* 2006. Instead, they are usually much smaller failures of slopes, in many cases the result of concentrated flows of surface water washing materials onto the track.

Although water can, and does, accumulate in the ballast underneath the rails, and may therefore be transported to a location on an embankment, it tends not to issue in such a way as to cause surface water erosion, although it may cause internal erosion as it exits, or infiltrate further to create a conventional 'landslip'.

CONCENTRATED WATER FLOWS RUNNING DOWN A CUT SLOPE – WHERE DOES THE WATER COME FROM?

The water running over a cut slope needs to come from somewhere, and for many infrastructure cut slopes it is not the precipitation that lands directly and immediately on the slope itself, because those 'engineered' slopes are generally quite small, and their steepness encourages run-off not to be concentrated but to happen as and when the rain arrives or snow melt happens. Instead, concentrated flows normally occur where the water is concentrated off-site and upslope.

The term 'precipitation' in the following includes rainwater and meltwater. Small amounts of precipitation may well simply infiltrate into the ground surface or even be held by leaves and other vegetation, thereby not even having any potential to infiltrate. However, there must be some rate of precipitation that exceeds the capacity of the vegetation cover to hold it, and then the surplus (which may be most of it) can then reach the ground surface in the catchment. What happens next is in part due to the ability or not of the water to infiltrate into the ground. Any water that does infiltrate, of course, contributes to the overall seepage pattern inside the slope, and that may be bad enough (as in the previous chapter), but what cannot infiltrate is available to form ponds and, in some cases, to spill over the slope. Fairly obviously, the most critical cases for surface water flows arise when the available water exceeds the ability of the ground to absorb much of it, and there may be a variety of scenarios in which this case occurs.

One of the cases is when the capacity of the ground to absorb water is reduced by antecedent rainfall conditions, but equally, a long period of antecedent dry weather may also affect that infiltration capacity of the ground. Ground that is frozen has a negligible infiltration capacity, and therefore rainfall or snowmelt happening before widespread thawing inside the soil mass has occurred usually gives rise to surface water flows. All of these cases have been blamed in the RAIB reports.[3] A particularly pernicious case was one in which, at the time of construction of the railway, a stream had been diverted across a cutting in an aqueduct. However, a century or so later it was recognised that the aqueduct's capacity did not necessarily meet the stream flow requirements, so a holding pond was constructed. As the holding pond was expected to discharge into the aqueduct it did not have a spillway, so that when an exceptional flow arrived at a time when the pond was already full, it had nowhere to go except to overflow down the railway cutting.[4]

Sidelong ground usually has its own natural drainage patterns that infrastructure earthworks may disrupt, and then it is important to be able to divert flows from any natural channels so that they do not pass over slopes steepened by construction. The natural drainage pattern may well have developed in response to commonplace, low-magnitude, precipitation events, or sometimes to much larger but relatively infrequent events, and so the fact that a natural channel of gully is not observed to carry water during rainfall while a site is being monitored does not mean that it will not carry extreme discharges in a different kind of event, and one should beware of any sort of drainage feature that appears to be dry, particularly if it contains vegetation that relies on wet conditions for part of its growing season. Some apparently dry channels only carry water from an ephemeral spring (called a 'bourne' in southern English dialects), driven at times of high groundwater levels. Sometimes streams seem to disappear into a sinkhole (or 'dene'), but one cannot rely on the discharge capacity of that feature to always work.

Surface water overflows are also sometimes the result of urbanisation, with areas such as roofs, roads, paved car parks, *etc.*, increasing the surface area of land which is effectively impermeable. Many infrastructure routes running through what were once rural areas are subsequently surrounded by urbanisation or even simply changes in land use which result in more runoff.

Blocked drains, or drains where the outlet has insufficient capacity, compound the problem, and indeed, where surface water is (in)conveniently disposed of in a 'soakaway' it may emerge somewhere undesirable, rather like an anthropogenically induced 'bourne'.

EARTH DAM STABILITY

I do not entertain the thought for a single minute that the subject of dam and reservoir engineering can be covered in a single chapter,[5] nor that even if it could, it would be right to do so. My intentions here are just to point out some of the slope engineering concerns that apply in the field of dam construction and operation. There are numerous instances of dams that failed during construction, and many of them were earthfill dams. Quite simply, if they were not stable under construction then they clearly were not going to function properly in the long run, even if by some miracle they had managed to be completed. Then, there are dams which were successfully completed but which failed subsequently due to slope instability brought on usually by some factor to do with the retained water, including overtopping, seepage finding a path through the abutments or even through the dam itself or under it, or some failures taking place around the reservoir perimeter.

I live in hope that some dam designer will feel motivated to write up their own experiences in a format similar to this one in a volume entitled '*Reflections on Dam and Reservoir Engineering*'. If such a book appears, then it is bound to have chapters that introduce the reader to the issues of hydrology and reservoir yield, whether or not the dam is provided for water supply, hydroelectric generation or to control floods. There will be chapters on the selection of different dam types and the investigation of foundation conditions. Almost certainly there will be chapters on spillways and draw-off structures. There may even be a chapter on the stability of earth- and rock-fill dams under construction and in operation, with hopefully a cross-reference in an end note to this book. Hopefully, its author will have read earlier works on the general subject.[6]

Quite a number of the developments in slope stability analysis were made in response to issues of earth dam stability. Not least was Bishop's work on the slip circle. Essentially, earth dams are particularly suited to wide valleys where the foundation conditions are poor and unable to bear the concentrated loads of masonry or concrete dams. This might lead one to suppose that the stability of the earth dam was greatly controlled by the properties of its foundation, but materials that weak would almost certainly be removed, and so the foundation had at least the strength of the earthfill and probably more. This situation contrasts with for example highway and railway embankments which are just built mainly on whatever the site presents, especially those of great age where the engineers involved either knew no better or were satisfied that for a small embankment the foundation would be adequate.

The problem of earth dam stability can be divided into the two sides of the dam: upstream and downstream. The downstream stability is very little different from any other slope except probably that the pore water pressures in it are controlled and hopefully much lower than in a natural slope or any other sort of earthworks and in general are dominated by the properties of the earthfill and not the foundation, although one should be aware that there is usually an issue related to the nature of the contact between the in-situ foundation soils and the fill. Dams may contain their own 'structural geology', created by different zones of fill, either determined by the available sources of material or to meet specific design requirements. Really good control over the placement of fill and its compaction will lead to comparatively uniform materials, although their properties may well have a sort of pseudo-stratification. Good compaction allied with careful treatment of fill after rainfall, for example by stripping off any wetted material before placing new, is useful to control settlement (although the control is never absolute), and also this is done to improve the shear strength properties of the contact. The stability during construction is, to a large extent, controlled by pore water pressures that are set up during the placement and compaction of fill including the changes in behaviour as the burial depth increases. Good materials for constructing the shoulders of earth dams are those that are dense when compacted and also relatively permeable, so that high positive pore water pressures are not generated or dissipate quickly. However, sometimes the shoulders will need to be constructed with clays or mudstones, and then drainage blankets may be required to dissipate construction-induced pore water pressures and appropriate slope gradients adopted.

At the originally constructed embankment for the Carsington dam, the shoulder drains were made from Carboniferous limestone, and the bulk fill was a Carboniferous pyritic mudstone: the pyrites oxidised and the sulphate ions attacked the limestone, generating carbon dioxide, which collected in a sump, eventually overcoming and killing four workers on the site. Geochemistry isn't often taught to civil engineering undergraduates, but since not all soils and rocks are chemically inert, such disasters can happen, although thankfully they are infrequent.

Where clays are used to form an impermeable barrier or core inside an earth dam it is likely that higher pore water pressures will be generated within them during construction, and therefore the stability of surfaces which pass through or into the core must take this into account.

During construction the upstream shoulder of an earthfill dam behaves very much like the downstream shoulder, but once finished, the stability behaviour is completely different because of the different water pressures engendered by reservoir filling and also by reservoir emptying when that is done. The stability of partly inundated slopes relative to them standing in the open air is affected by two broad factors discussed in the previous chapter, but revisited here because the matter is so important. Firstly, the external load provided by the weight of water *increases* the stresses in the upstream shoulder and also through the dam, and their effect generally increases stability. Secondly, and fighting against the support provided by the external water, the presence of external water may increase the pore water pressures inside the dam and reduce stability. The two effects battle each other, and generally the pore water pressure increase and consequent reduction in stability beats the supporting effect, with the worst combination appearing at some level of submergence that is called the 'critical pool level'. While the critical pool level can be computed by undertaking stability analyses at different levels of submergence, it is always a good idea to fill the reservoir for the first time rather slowly and monitor what happens inside the embankment.

A worse combination is when the supporting effect of the external water is withdrawn while leaving substantial internal pore water pressures in the upstream shoulder of the dam. This case is called 'drawdown', and it may be particularly critical when the drawdown is rapid, because the support is withdrawn quickly, and the pore water does not have time to escape. Examples of drawdown failure are known not only in the upstream shoulders of earth dams but also around the periphery of reservoirs in the natural slopes. A particular example or set of examples occurred around the periphery of the reservoir of the Mohne Dam breached by 'bouncing bombs' in the famous 'Dambusters Raid' of 1942: the dam itself being of concrete construction and therefore not susceptible to the effects of rapid drawdown, whereas the reservoir side slopes were in their natural state of soil slopes in the valley prior to impounding.

Since the critical pool level on drawdown may be lower than the critical pool level on filling, the immediate knee-jerk reaction to an observation of movements in the upstream shoulder of an earth dam during first filling, which is to lower the water levels again, may not always be sensible even if it is the only recourse available.

Drainage blankets incorporated in the upstream shoulder during construction may prove of great benefit in allowing the outward drainage of water when the reservoir level is drawn down.

ROCKFILL DAMS

Rockfill dams have the advantages, as far as stability is concerned, of being constructable to steep slopes (which reduces the volume of materials needed) and of being relatively permeable. In combination with asphaltic or concrete narrow vertical cores, rockfill can be a very economical solution. On the downstream shoulder, pore pressures are usually low to non-existent, and the angle of shearing resistance is high, so stability is less of an issue than with earthfills. Similarly, on the upstream shoulder the rockfill drains rapidly on drawdown with equivalent positive behaviour. Clay cores are less useful in rockfill dams because of issues relating to differential settlement and the provision of filters to control internal erosion. Clay cores always need careful construction of filter zones, especially where the adjacent shoulder is highly permeable.

WHEN IS A DAM NOT A DAM (OR *VICE VERSA* IF YOU PLEASE)

The case of an embankment with a high water level on the upstream side is pretty obviously a dam, whether or not it is called one. There are numerous cases where some of the same stability issues that define embankment dam engineering come into play, and that is even when one excludes the case of a landslide that is dropped into a river valley and makes a natural obstacle to flow.

One of those other cases is a canal being carried on the crest of an embankment or being retained by an embankment on one side along sidelong ground. Even more so than railways, canals necessarily follow topographic contours with inevitably long lengths of water-retaining embankments on their downslope sides. It is not always obvious just how much water is retained, because that is a function of the length of canal between points such as locks that can compartmentalise the water, rather than it is related to either its width or depth.

The limited speed of canal boats means that they are less vulnerable to running into debris than are road vehicles or trains, but the wash from boats and their propellers is fundamentally erosive, especially where the canal route changes direction. In eroding locations some protection is required, commonly in the form of sheet piling which has the additional beneficial property of controlling seepage through the bank.

Surface water erosion with flows from an impounded water source (reservoir, canal, storm water, detention pond, *etc.*) results from lack of freeboard. In turn, the inadequacy of freeboard may result from such causes as increased water levels due to such factors as the failure of spillways, extreme floodwater levels or, on the coast, wind-driven runup in addition to high tide, or to lowered embankment crest levels due to settlement or subsidence. It is not always necessary for water to come over the top of an embankment if it can find an easy seepage path through animal burrows or soil desiccation (shrinkage) cracks, or defects in a core or other waterproofing measure.

Another case where there may be a large quantity of water on one side of a long embankment is the case of river training embankments, originally constructed to prevent low lying land from inundation at times of high river discharge, or perhaps to realign a meandering channel to something more convenient for the adjacent land use. Particular examples abound, and nowhere more notably than in the case of the embankments constructed along the Mississippi River through New Orleans that failed during Hurricane Katrina on 23rd August 2005. Lengths of the embankments had been raised with walls (referred to as I-walls) at their crests, as such a measure did not require the additional land take that a conventional increase in height of an embankment would entail. Overtopping an I-wall gives the flow more erosive power than overtopping a simple embankment – bad enough in itself – and the forces on the wall also affect the stability of the underlying slope. The flow is continuous, and its duration is longer than a wave overtopping the wave wall on a rubble mound breakwater, but the I-wall of the Mississippi flood levees is otherwise analogous to a wave wall.

A problem with river training embankments arises when the water level in the river is more than anticipated in design, compounded by the practice of building them from homogenous fill, usually with a high clay fraction and without engineered internal drainage. The foundations of river training embankments, especially those of any age, are inevitably of somewhat variable quality.

In the UK's peatlands, rivers may well flow at higher elevations than the surrounding ground between training embankments because the ground level generally has reduced due to peat shrinkage or worse, by wastage when the peat has decomposed when man has lowered the groundwater level to make the land more accessible. In both river-training and coastal defence embankments, layers of gravel at the base of the foundation soils where they overly the solid geology may lead to hydraulic failure as the high water pressures on one side of the embankment are propagated underneath it.

An analogous problem to that of the river training embankment is the flood defence embankment used to reclaim coastal marshland, and often constructed as a homogeneous fill embankment from the marsh deposits. It would be nice to think that the Dutch experience of land reclamation using flood defence embankments of considerable size involved more engineering than the rather low embankments often found in SE England, and in many respects they are very similar to river training works. The storm surge leading to unprecedented high tidal levels in the North Sea Flood of 1st February 1953 overtopped many flood defence works and eroded large sections, leading to multiple fatalities in the UK and worse in the Netherlands, necessitating the construction of better and higher protection works afterwards, and in some places, surge barriers in estuaries.

Perhaps less obvious is the case where there is free water carried to the crest of a slope by movement along the ballast of a railway or along a pea gravel bedding for a pipeline or other services. The pipeline case is exacerbated if the pipeline leaks, leading to high fluid pressures or pollution of some kind and possibly even artesian conditions which are damaging to stability wherever they occur in a slope.

High water pressures at the crest of the slope also arise where there are badly positioned soakaways, septic tank drainage fields, leaky reservoirs or swimming pools, or 'sustainable urban drainage systems' (SUDS).

SPILLWAYS AND OVERTOPPING

It is always a bad idea to allow any water-retaining earthwork to overtop, although overtopping is more likely with any form of reservoir that *impounds* than a reservoir that *stores* and contains water actually pumped from somewhere else, as you can always stop pumping! Impounding reservoirs always carry the risk that heavy river flows coincide with reservoir-full conditions, and then the surplus water must be discharged in some way, ideally *via* a spillway that has adequate capacity. Even when a spillway is provided, there comes the problem of what happens in the river channel downstream, and even an ostensibly well-designed scheme that carries excess water successfully through a number of wet seasons may underperform if the necessary discharges are more than allowed for in design, the spillway or stilling basin suffer from lack of maintenance or something else goes wrong.

In one case I saw, the spillway capacity had been increased by providing 'hoods' so that the overflow spillways turned into symphonic spillways, and the resulting increased discharge rates led to scour immediately downstream of the masonry dam. The scour undermined adjacent soil slopes threatening a road, and also cut into the valley floor scouring out some dykes that were weaker than the country rock. The dykes were not recognised earlier, but were the reason why the valley had formed where it did in the first place – a lesson that the landscape forms where it does for good geological reasons.

A case I followed on YouTube was that of the failed Oroville Dam spillway in California (February 2017). In that case even the emergency spillway overtopping caused huge slope erosion, and the main spillway equally deep scour, for which the word 'gully' is a poor description of the scale. The materials removed were, of course, deposited downstream.

An even more dramatic event occurs when a dam is breached, and I note that there are numerous books as well as papers on the subject. Interesting though the topic is, I suspect most dam engineers work hard on making sure that their designs are never breached, but there are many old dams designed and constructed before the development of modern geotechnical science, and they are (usually, but not always) at more risk of developing problems.

There are a few cases of dams being overtopped by waves generated by landslides (or glacial collapses into the reservoir), probably most notably being the Vaiont landslide and flood of October 1963. The huge wave rushed down the Vaiont Gorge, and scoured a big hole in the sediments in the bed of the River Piave, compounding the effects on the town of Longarone. Unusual flows inevitably disturb sediment accumulations which are in equilibrium with ordinary flows.

TAILINGS DAM BREACHES

Dam breaks for water-retaining dams usually give rise to floods downstream, but after all, it is 'only' water (unless it sweeps up sediment on the way, which it so often does). Tailings dam breaches release the tailings, which start off containing a great deal of solids, and depending on how they were produced, all manner of pollutants, and thus when their floods subside, they leave more sediment behind. Generally speaking, inadvertent releases from tailings dams arise not only from initially unstable retention bunds, but also from overtopping and the resultant scour, especially if the tailings lagoon was constructed in a water channel, as so often they are.

VEGETATION AND SURFACE WATER

Vegetation is often hailed as the solution to slope instability because there are numerous cases where tree clearance has resulted in a slope that being exposed becomes subject to shallow instability. However, the converse, which is to plant new vegetation, may take a considerable time for the vegetation to become established and is therefore, even in the best case, not an immediate solution. Vegetation has at least three effects, the first of which is to capture precipitation before it even lands on the ground and thereby enabling the water to evaporate off and not contribute in any way to the pore water pressures in the ground. The second effect is through evapotranspiration to actually remove some of the water that was otherwise in the ground. A third effect is to control erosion when surface water passes over the slope. All three of these mechanisms can only be considered to be second-order effects relative to soundly constructed engineered works.

Vegetation is not without its downsides. Deciduous trees shed leaves in the autumn, leading to the blockage of drains or even causing a skid risk on a road or loss of traction on a railway (as evidenced by the notorious 'Wrong kind of leaves' quote), and once leaves have been shed then deciduous trees contribute little to evapotranspiration. Coniferous trees in any case are less effective at removing water from the ground, often because of shallow root systems but also from the nature of their leaves. Fallen trees or even just fallen branches may cause problems to infrastructure routes, and a dense tree cover in any case restricts access to a slope for inspection purposes.

UNUSUAL PORE FLUIDS

This section is here because the chapter is short, and I couldn't find anywhere else that it actually fitted! Freshwater has a lower density than seawater, but the water in the Dead Sea is denser still. Petroleum products are lighter than water, and gasses even lighter. All are possible pore fluids and not just on their own but in combination. I mentioned these particular fluids because sometimes we are faced with those non-aqueous pore fluids. The 'hydraulic' properties of petroleum products became an issue in the design of containment bunds built to capture leakage from oil tanks. Oil, of course, has a different viscosity to water, and this property affects the hydraulic conductivity (permeability) of a soil. Moreover, the different surface tension behaviour means that the 'wetting front' advances at different rates with different fluids and, in any case, is different with different degrees of water saturation of the bund fill.

Gasses, such as municipal solid waste derived methane or volcanic gasses, also differ greatly in their behaviour relative to freshwater pore fluid. Many gasses will go into solution in some pore fluids depending on pressure, and therefore the degree of saturation of the soil mass may change under changes of total stress (loading or unloading). The phenomenon is well known in soil testing and leads to the necessity for a saturation stage in the triaxial test, whereas in-situ, it results in changes to the pore pressure parameter B and hence also to $\overline{B}$.

Some soil gasses are flammable, and that, for example, means that it is advisable to avoid naked flames or indeed any source of ignition when visiting a failure in a municipal waste storage facility as well as when obviously flammable fluids are present.

Finally, some pore fluids will dissolve chemicals in the soil or otherwise react with them, with many salts being particularly soluble in water, or cementitious minerals like calcite being susceptible to attack by weak acids, and the dissolution properties thereby causing changes in soil strength parameters.

The point of all this is, of course, that the behaviour of soil is complicated and only reasonably well understood in relatively simple situations. It is possible to model soil behaviour numerically even in the partly saturated case, but in my experience the saturated case is usually critical. In words of one syllable, metaphorically speaking, relying on suction in the soil mass to maintain stability is also relying on it never raining!

SUFFUSION, SUFFOSION – INTERNAL EROSION

As an afterthought of sorts, and something that may well have been appropriate in the previous chapter, there is the matter of internal erosion, or the removal of soil particles from the interior of a slope by the outward flow of water. Most engineers will be aware of the process by which surface water flowing over a slope can pick up and transport soil particles of various sizes, but unless they specialise in dam engineering, they may not have given thought to *internal erosion*. Indeed, the flow of water through many soil masses is far too slow to transport anything due to the low hydraulic conductivity in many soil masses. In some situations, however, the flow of water may be sufficient to dislodge small particles, and when that leads to the enlargement of the soil pores, the hydraulic conductivity increases, the flows become larger and faster, and the whole process may ultimately lead to the collapse of the slope and the release of the source of the water.

In soils where the small particles that partly fill the void space between a network of larger particles are small enough to pass through the apertures between the larger particles, the process is sometimes referred to as *suffusion*. (Soils in which the large particles are isolated from each other such as random boulders in a much finer grained matrix do not operate in the same way.) Suffusion is assisted if there are cracks or other open discontinuities in the soil mass assisting the flow of water. A related process occurs where the soil is dispersive, that is where just simply wetting causes the soil particles to disaggregate. You can demonstrate dispersivity by dropping a dried piece of clay into a container of water, but some dispersive soils react the same way even if they are already wet.

Suffosion is probably a composite word linking *suffusion* and *erosion*, which is where water flowing through a coarse-grained soil removes fines from an adjacent finer-grained soil, and just as the word appears to be a composite between two other words, the process is also a sort of halfway house.

Mostly, internal erosion as a catch-all term covers both suffusion and suffosion, and is likely to be of greatest concern to slope engineers working on water-retaining structures such as dams, mine waste lagoons or canal banks, although occasionally one finds internal erosion occurring in a silt or sand bed in a sedimentary sequence. Sometimes the erosion leads to the formation of pipes through the soil, and is therefore also known as 'piping'.

Internal erosion is best controlled by the insertion of filters into an earthwork to prevent the escape of those fine particles from a particular soil zone (such as a dam core) into an adjacent soil zone (such as a transition zone or the shoulder of a dam). It is much more difficult to prevent the removal of fines where water exits from the face of a slope, and may require not just the filter or filter layer, but also a soil berm to keep the filter in place. Filters may be, in some cases, geofabrics; in others they are soil layers with intermediate grading between the soil to be protected and the adjacent more permeable zone.

NOTES

1 The RAIB as a Government department maintains an excellent website with both summaries of its reports and the full reports themselves. The claim is made that it does not seek to attribute blame, but as in some cases the participant

organizations are named and in others they are not, some doubt must remain as to whether or not blame is implicitly implied. If searching, be sure to use the term 'landslip' in preference to 'landslide', as the former is official railway parlance and has been since the earliest railway was built. The NTSB in the US covers other infrastructure issues including aviation and pipelines.

2 *The Railway Children* by E. Nesbit, published in 1905 and dramatised numerous times is a perennial favourite of readers of all ages. Both the landslide and the subsequent remedial works are described *en passant* as observed by the child protagonists.
3 See RAIB reports Nos11/2017 and 10/2018. Some earlier reports are not numbered and are titled only with the location. A search of the RAIB archive, despite its short period of coverage, is highly instructive.
4 See RAIB report No 04/2020.
5 If anyone doubts my view on this then I refer them to the excellent book: Fell, R., MacGregor, P., Stapledon, D., Bell, G. & Foster, M. (2015) *Geotechnical Engineering of Dams*. Second Edition. Routledge.
6 Including, I hope, Dam Geology by the fabulously named Rupert Cavendish Skyring Walters (1st edition 1962, enlarged 2nd edition 1971) published by Butterworths, or perhaps Coxon's report on the failure of the Carsington Dam, which contains more facts than some of the cover-ups published subsequently. See Coxon, R. E. (1986) *Failure of Carsington Embankment: Report to the Secretary of State for the Environment*. HMSO. Another good read is Saxena, K. R. & Sharma, V. K. (2004) *Dams: Incidents and Accidents*. CRC Press, even though the authors sometimes lack the inner truth of the failures that they report.

Chapter 8

Dealing with the landslide problem without stabilisation

INTRODUCTION

There are numerous things that a person or community faced with a landslide problem can do in response that don't involve stabilisation. These options can be considered in the light of risk reduction, so that if you consider what constitutes risk, you have the spectrum of activities that can reduce it.

Risk is essentially the potential for loss, and we consider that the potential loss in a particular time period, say a year, relates to the probability that a damaging or loss-causing event happens in that year. It also depends on the number and value of elements at risk as well as the proportion of their value that might be lost, ranging from total loss to some form of repairable damage. There are different types of damaging events that are generically categorised as *natural hazards*, but here we are concerned solely with landslides. We should note that there are multiple elements at risk that might have different vulnerabilities to various types of landslide, so the true value of risk has to summed up over all the types and probabilities for the events with the consequences in terms of loss for each. Even so, the computation of the likely loss per annum may be so small that there is little justification for an expensive intervention, and the likely landslide types and magnitudes considered tend to be those that have been experienced in the recent past.

My mental picture for this is a rock slope somewhere in a sparsely populated mountainous region but threatening principally a rural road and some fields in which a small number of farm animals are kept. Clearly, there are some elements at risk that are static: the road itself, fences, and so on, but there are movable things: vehicles, hikers, the school bus, tourist coaches and, indeed, the animals. It then comes down to the magnitude–frequency relationship in the rock falls, the likely impacts on not just the elements themselves, but also the knock-on effects such as cars being scraped, animals released through broken fences, or even near misses that give people a fright or other intangible losses and damage. In the end, it is probable that in this scenario the risk amounts to a sum that basically commands the simplest and cheapest solution that is second only to doing nothing and just living with the risk. All the strategies that do not involve stabilising the rock face

DOI: 10.1201/9781003428169-8

are encompassed in this chapter. Risk analysis is a complex problem, and it demands that a book on the subject is in your library, probably the one by Lee and Jones.[1]

One of the strategies is to move elements at risk somewhere else where there isn't a landslide problem. In my little conceptual case it might simply be a matter of moving the road away from the foot of that rock face, but presumably land in the valley is earmarked for other purposes, and who knows – there may be less stable rock faces on the other side of the valley. Even if the land is available, it may come with a different set of hazards.

We see the relocation strategy on a grand scale in southern Italy,[2] where several hilltop towns affected or threatened by landsliding have simply been deserted. These include Campomaggiore Vecchio and Craco, together with many more deserted after earthquakes and massacres by the Napoleonic French and the Nazi Germans, the two abandoned after landsliding being a relatively short drive from each other down the Autostrada. In the case of Craco the remedial measures that were constructed failed to solve the problem and the community gave up, but in Campomaggiore Vecchio the community moved to a new location a couple of kilometres away. The new town of Campomaggiore is also now threatened with the encroachment of landslides, but the old one was deserted over a century ago, and Craco more than half a century ago, so these (and many other) landslide problems in the south of Italy cannot be the result of anthropogenic global warming or climate change or whatever it is called these days. Instead it is much more likely to be the result of delivering a water supply and disposing of the waste water inappropriately.

No matter how advisable it might be to move somewhere not subject to landsliding, the alternatives sometimes contain that different but equally undesirable set of hazards of their own. For instance, moving from a hillside to the flat ground at its foot may take the element at risk away from the landslide, but puts it at risk from flooding. And some risks simply cannot be avoided without a really big relocation, because a whole country can be subject to more or less the same seismicity, although landslide risk is unevenly distributed. In the past, the high ground was defensible, and while this may not be a consideration today, it does account sometimes for why communities are located where they are.

Anyway, moving elements at risk to even a slightly safer location reduces risk, and that may be enough. The road that climbs up from Chesil Beach to the heights of the Isle of Portland in southern England was realigned only a small amount to avoid cliff-edge instability at Priory Corner,[3] and that was enough to radically alter the damage to the road routinely experienced and stands as a good example of the approach. It won't work on a geological timescale, but it will work on the timescale common to most civil engineering works and probably longer.

It is also possible to reduce vulnerability to landslides, or the expected proportion of the value of those elements lost in a landslide event. For example, a building in an area subject to small rockfalls from an adjacent cliff might have reinforced or stronger walls on the uphill side, or the most occupied rooms laid out with a downslope aspect on the side away from the hazard. Of course, there could be an event so large that it destroys the whole building, but the relative probability of that must be smaller than less significant events.

The commonest response to natural hazards seems to be to ignore them until they occur and then to complain loudly that the government should do something. I have to admit that sometimes I lack sympathy, as I have none for protestors who complain about airport noise when the airport was there long before they bought their house. Complainants who live, by choice, in an area subject to coastal erosion or on a known landslide likewise have to get on with it. There is an all-too-frequent desire to socialise the costs of dealing with natural hazards but to privatise the benefits among a small community. Small communities, however, can rarely afford the necessary costs of a remedial scheme without financial help from outside. Life can be hard sometimes.

The sympathy comes, of course, when the actions of others exacerbate the problem, for example, dredging for a container port or construction of a harbour arm causes increased erosion downdrift. In this case the liability should accompany the causation, and calls for appropriate rectification or compensation need to be answered appropriately as part of the project economics.

In some instances, not just the perception of risk but the actuality might be sufficiently small to be tolerable, but there may still be a case to reduce that risk even further. For example, the construction of a netting rockfall barrier[4] between a building and a cliff may not be strictly necessary, but it may provide peace of mind to its occupiers in a way that is difficult to monetise but is valuable nonetheless.

In mountainous regions, or indeed anywhere where there are rock cuts, the probability of vehicles being hit by falling rocks is less than of driving into a rock, so it is useful to improve driver awareness by the use of hazard warning signs and speed limits, or in urban areas, to provide street lighting. Another important aspect of driver awareness includes knowledge of what to do if a road-blocking event is encountered. My advice is not to remain stationary in the vicinity where secondary failures might occur, but to get away as quickly as possible to a safe distance. This might not help get the YouTube 'prize' for the best video, but it may stop it being a posthumous award.

Railways pose a not-so-subtle different exposure to hazard from roads, because even though the driver may be much more attuned to the problem, there is often less that can be done to avoid it, as trains cannot swerve, nor can they stop quickly; and in addition they may be so long that even if the event misses the engine and driver, it may hit farther back in the train, whether or not it derails it. In all sorts of transport infrastructure, landslides may block, damage or destroy the permanent way, and its effect is then projected well beyond the footprint of the event, although usually *via* its economic and social disruption. In the case of landslides that affect canals, reservoirs, and dams, the release of water may project physical hazards well beyond the footprint of the landslide *per se*.[5] The same proviso also relates to landslides into reservoirs, which may create an intense but short duration flood by overtopping.

In addition to static warning signs, systems can be deployed that automatically register landslide activity and light 'stop' signs. A simple system deployed in the railway between Dover and Folkestone across *inter alia* the notorious Folkestone Warren landslides is simply a wire with easily separated connections at intervals. A falling rock that breaks the connections would set the signals to 'danger'. Other systems based on laser beams or radar can be used, but the more complicated the technology, the greater the risk of malfunction and the greater the need for regular routine maintenance and testing. False alarms are a problem for all warning systems but range from simple inconvenience where the discipline of responding is enforced (as it is with railways, for example, where passing a signal set to danger is a serious offence on the part of the driver) and may generate a false complacency, as with dynamic speed limits on motorways, where false alarms create dangerous behaviour in motorists who convince themselves that the signage is always misleading.

A fundamental problem with the perception of risk is that it is rarely based on evidence, occasionally based on expert opinion, but much more often based on emotion. This is most true when the risks are not obvious to lay people, often because their long-term experience has been either event-free or of a different kind. For example, in the village of Casso, over 200m above the water level of the Vaiont reservoir, but threatened with rock and debris slides from the mountain above, it must have been inconceivable that a landslide would propel a wave up a cliff to suck 27 people out of the village to their death, and even professionals at the time estimated that the landslide into the drawn-down reservoir would be much smaller and so took up position near the dam crest to observe it. This was not simply an expert *opinion*, it was based on physical modelling that proved wrong when put to the test. The inconceivability of the form of the disaster must be a lesson in quantitative risk analysis, and that is just how difficult it is to predict what has never happened before.

In my own experience, I was once accosted by a man bearing a knife whose view – largely because what I had said to the press had been misreported and sensationalised[6] – was that I had affected the value of his house, one of the last few remaining in a street devastated by landslide movement. A few months later, he was nowhere to be seen, and neither was his house. It had fallen victim to those self-same ground movements and was demolished as an unsafe structure. I deeply regretted that I missed the opportunity to sympathise. Or perhaps not.

A particular response to slow ground movements associated with landslide activity is to construct buildings on jackable frames so that they may be relevelled from time to time. This gives rise to issues with connections to some services and creates an undercroft that can harbour vermin or attract small children. On a steeper slope, the undercroft can alternatively provide good storage space. In one example, the frame was not stiff enough, either as a design error or due to modifications to the building and heavier-than-envisaged contents, so it sagged and caused damage through its deflections, so this is not a foolproof (or sometimes futureproof) design approach. In another bizarre case, a planning inspector refused a planning application to replace a house that had been ripped apart by landslide movements with a timber structure supported on a jackable base on the grounds that he was bound by Planning Policy Guidance (then, it was an early version of the guidance note PPG 14) not to permit building on unstable ground. The equally bizarre corollary was that the inspector was perfectly content for the original house to be reconstructed in traditional materials. Sometimes it seems one has to work with – or against – the complete imbeciles who infest all walks of life.

On occasion, landslide risk reduction measures can be incorporated sensibly in building regulations and statutes. A good example of this occurs at Sandgate in the Folkestone District in southern England, where a set of regulations apply to development on the Sandgate landslide complex. The regulations are known as the Latchgate Conditions, named after a block of apartments, where the regulations were first developed and applied. They require developers to conduct a ground investigation and a stability assessment to show, at the very least, that the site stability is not worsened. Developers consider such regulations unduly restrictive and tend to try to counter or avoid them – after all, developers often have recourse to liquidation and often are not around when problems arise. In my experience, contractors are also fond of disappearing and reappearing under a different guise, or contractors' insurers pursue designers' professional indemnity insurers with what can be sometimes described as claims without a great deal of merit.

As in the case of the PPG 14 planning refusal, sometimes the application of the regulations is simply eccentric. I well remember a case where a UK firm had bought a development site and company in the USA. The site was occupied by a set of deep-seated landslides, and these had been investigated by means of large-diameter shafts entered by geologists who logged the succession, including the location of slip surfaces. I was involved in developing systems to stabilise these landslides to allow development to proceed in safety. However, the plug was pulled on the whole investigation and design process from the US end when it was realised that the local building ordinances only required a stability assessment if the dip of the beds exceeded a certain value. At this site the dips were sub-horizontal. Now, approaching 35 years or so later, I sometimes look on the Internet to see the landslide damage that has occurred. There is plenty.

The disposal of water can be a problem, with many engineers favouring soakaways or 'sustainable drainage', perhaps in response to being forbidden from discharging land drainage water into streams or rivers. In areas of particular landslide susceptibility, adding water to the ground is certainly anything but beneficial. This may be the case with surface water runoff from roofs and highways, but it can be worse with leakage from swimming pools where the chemistry of the pool water attacks weak calcareous rocks (such as the Chalk) or calcareous soils; or the water from septic tank drainage which adds volume to chemistry and contaminates the ground with coliform bacteria. In small quantities, natural soil bacteria can beat the residue of the human gut, but large volumes of contaminated water overwhelm the soil bacteria. This does mean that constructing a drainage system for stabilisation in such localities must be done with due regard to the health of the workforce, and the discharge from the system may need to be treated before it can be disposed of in an environmentally friendly manner.

A particularly pernicious case that overloads septic tank drainage is to proliferate individual installations in a settlement, or to take a large building and convert it for multiple occupancy. Widely spaced residential developments may be accommodated with septic tanks without necessarily representing a major impact on groundwater levels, but any quasi urbanisation in a sensitive location should be associated with connection to a mains sewerage system. Polluted groundwater also leads to the abandonment of wells, which may have had a beneficial effect on groundwater levels when used for water abstraction. Householders may have very little understanding of the volumes of water they discard, as it is often unmetered. The volumes used for lavatory flushing may be small in comparison to bathing but contaminate everything, and out-of-sight uses like plumbed-in washing machines and dishwashers may consume and discard large volumes without the user being aware of how much is used. Garden watering or irrigation may also be a big water user. In addition, water supply and disposal systems are notorious for leakage, especially where their joints are loosened or pipes are broken by ground movements.[7]

BRIDGES OVER, TUNNELS UNDER AND PREFERABLY NOT THROUGH

In one case that I worked on, the client organisation preferred a bridge scheme under which the landslide could continue to move because it had in-house – but under-employed – structural engineers but no geotechnical engineers. The bridge (actually a viaduct) had piers in the landslide that were founded on bedrock below the landslide and constructed sufficiently strongly with a 'cutwater' planform so that the forces imposed by the landslide as it slid past could be safely accommodated.

More commonly, a bridge or viaduct may be built to span across a valley that periodically experiences debris flows. One hopes that the abutments are never undermined. An example in Hong Kong also poses the same question. This was the Po Shan Road landslide[8] that appears to owe its origins not only to the adverse geology and weathering, but also to water flows from a road high up on the slope. The road was reconstructed with a composite construction steel beam and concrete slab bridge across the landslide scar (which was unlikely to fail on a less than geological timescale as all the weathered material had already slid off).

One way of avoiding a landslide with a road, railway or canal is to tunnel around the landslide or underneath it in the solid. A tunnel also has the advantage of potentially intercepting groundwater flows, especially if drain arrays are drilled from it. You should take care with any tunnel into a landslide, because that usually means entering the toe, and that in turn may be a combination of unloading and removal of some part of the slip surface. A cutting in the approach to a tunnel portal also potentially unloads the toe of the landslide. Moreover, any tunnel has to cross the edges of a landslide and may therefore suffer a cracked lining with even the smallest of movements. Tunnels that carry water, such as diversion tunnels or headraces for hydroelectric power plants, may carry water under pressure that, leaking out through a cracked lining, may cause a landslide to move. All services that carry water have the potential for leakage too.

ROCK SLOPES

Many methods of dealing with rock slopes do not rely on stabilisation, which would largely mean preventing the slope from shedding any blocks at all, or certainly not big ones, but instead the non-stabilisation options consist of controlling the detachment of blocks from the face and reducing their velocity, capturing the falling blocks, or warning people in the vicinity of the slope of the hazard so that they can keep clear or adopt a behaviour that minimises the hazard.

A method of controlling the fall of small blocks is to provide steps or benches in the profile, each of which preferably has a catch fence. Benches usually need to be wide enough to allow vehicular access, as to make them narrower means that clearing debris periodically must be done with manual labour with all the risks that entails. Catch fences can also be installed during excavation of a rock slope, with the advantage of making safer working conditions at the bottom, but with the disadvantage of making the fences themselves more difficult to maintain or replace if damaged. An alternative method is to provide mesh draped over the face. There seem to be two philosophies about this, with one being to simply secure the mesh at the top and let blocks slither out in a controlled fashion, and the other is to secure the mesh over its entire extent with a pattern of rock bolts. Where I have seen the two, my preference is with the draped method, as the build-up of small rock blocks can eventually overwhelm the tensile strength of the mesh and cause it first to bulge and then to rupture.

The choices of corrosion protection for the mesh include galvanising or plastic coating. Black plastic coating makes the mesh harder to see from a distance when set against dark-coloured rocks, and this may be preferable in a setting where less visual intrusion is better. In either case, the mesh has a finite lifespan, and if used in permanent works, needs access to repair or replace.

A combination of steel mesh and sprayed concrete, held in place with rock bolts and provided with drains to prevent a build-up of water pressures behind the facing is sometimes used, especially where visual intrusion is not a problem. Where I have seen this, it was definitely not a stabilisation solution, as several rather large areas failed! The concrete and steel bolts in the debris add to the difficulty of clearing up after a failure.

At the foot of the slope, rocks may be contained with fences, barriers made from earth, rockfill or gabions, or with ditches (preferably with energy-absorbing gravel), or a combination of barriers and ditches. In particularly hazardous locations barriers can be built from steel sections founded on piles. Rock fall barriers probably need periodic 'emptying' if the accumulated collection of fallen rocks is not to provide a 'launching ramp' for other blocks.

Warning signs are common in areas of landslide hazard, not only at the foot of slopes but also to keep people away from the slope crest. One of the internationally recognised warning signs for motorists is that of a slope with rocks falling. I particularly dislike this sign when the highway authority has provided it in the wrong sense (where, say, the sign shows rocks falling from the left whereas the rockface is on the right), but it should encourage motorists to take care. These signs can be supplemented with speed limits to minimise the risk of a motorist hitting an already fallen rock. Statistically, with typical lengths of vehicles, the relative hazards of hitting or being hit by a fallen rock make the latter less likely, and in addition, vehicles are relatively good at accelerating, braking or swerving, which also helps, but only if the driver is alert. This is not the case for train drivers, who can do little to avert their train being hit by falling rocks even if the driver sees them ahead and may need subsequent treatment for stress or worse.

Street lighting is also helpful in aiding drivers to avoid obstructions in the road, but it may not be feasible to provide it in a rural location on grounds of cost, maintainability, and of 'light pollution'.

Pedestrians and animals may also be at risk, with fast-moving animals sensitive to ground vibrations or with excellent hearing less so than pedestrians whose attentions may be elsewhere, although farm animals may escape from fields if fences are demolished by rockfalls and animals may present a hazard to themselves and others if they are panicked.

SURFACE WATER INTERCEPTION FOR ROCK SLOPES

Drainage for rock slopes that removes water that is already in the ground is similar in most respects to drainage in soil slopes, and that is discussed in the next chapter, but many of the methods that fall into the generic category of surface treatments *via* drainage seem to fall here[9] naturally following on from the previous section. This is because drainage as a stabilisation measure stops water entering the ground or removes it when it is already there, but for rock slopes it is mainly to prevent water running down the face and contributing to exfoliation of small pieces of rock.

It is fairly common to provide a ditch round the periphery of a rock slope that channels the water that is collected round the sides of the rock cut or down existing channels. One of the side benefits of working in a hard rock area is, of course, the availability of stone to line channels, which has the added benefit of making them blend into the landscape more effectively than using concrete. Particularly with steep channel sections the roughness of a masonry lining to a channel dissipates some of the energy of the flowing water and helps aerate it, which can be a benefit in temperate upland areas where the water may be contaminated with animal droppings or peat.

Where a rock cut is associated with a road or railway on sidelong ground, the water must cross the route in a culvert or covered channel, and this normally requires a grating to prevent access by animals, people or trash, but all gratings need regular clearance if they are not to become blocked by an accumulation of vegetation, dead sheep, litter or other refuse.

Drainage measures also include collection and disposal of water that issues from permeable or semi-permeable strata, water that can sometimes cause internal erosion of closely jointed rock. In this case, the solutions akin to dentistry are used: the stratum or fault is raked out and the cavity filled with sacks of rock chippings or pea gravel, and the whole finished with a masonry face using the rock from the cutting excavation. The similarity with dentistry is recognised by calling this 'dental masonry' or 'dental work'. Masonry can also be used to support overhangs. The drainage media can become clogged in the way of most drains with precipitates of minerals such as iron compounds or limescale, and the exit for water is at risk in certain climates of freezing. The expansion of water when it freezes is the cause not only of damage to these structures but generally leads to the exfoliation of fragments from the surface, a phenomenon that also can occur in hotter climates with diurnal temperature changes even if they never freeze.

Occasionally the dental work is called a *chimney drain* from its resemblance to a rustic masonry chimney that is an architectural feature of some buildings, notably but not only those constructed from wood.[10] It is not the same as a chimney drain built inside a dam.

LIVING WITH THE LANDSLIDE RISK BUT WITH SOME ADVANCE WARNING

Ground movements in buildings and infrastructure are most easily accommodated if the deformations are small, or if larger, then they are regularly experienced and therefore accepted.

Mostly, however, a community lives with a landslide threat rather than continual landslide activity. The collection of records in places like Hong Kong means that actual rainfall can be recorded, and if it approaches critical levels determined from experience, then warnings can be issued. Where the records do not allow such precision, only generic warnings can be issued, based on predicted rainfall *via* forecasts, radar, and similar means. Tropical storms also bring heavy rainfall, and these can be observed in advance by satellite or 'hurricane chaser' aircraft. Snowfall generates the risk of snow landslides or avalanches, sometimes containing rock, a topic that I cannot add much to, except to note that artillery[11] is sometimes used to precipitate small avalanches and that snow mechanics has an overlap with some parts of soil mechanics. For slope instability, snow melt is more critical, as apart from ice wedging, the pressures from water at depth are generally more significant than the weight of snow at the surface. Moreover, melting snow keeps the underlying ground surface wet (if it is not frozen, of course) and inhibits runoff, both factors increasing the potential for infiltration.

Landslide hazard warning systems may make use of instruments in the body of particular landslides, such as the piezometers and inclinometers used in ground investigations, complemented with surface movement observations, which can be divided into two categories: surface movements on the landslide itself, and observations of what happens in the landslide track.

Surface movements can be detected using topographic survey methods, but unless a robotic instrument is used the frequency of survey measurements is too limited to detect the onset of many types of movement. GPS is rather too expensive to put many observation stations on a landslide. Remote sensing with laser scanning, INSAR, satellite observation, *etc.*, are all useful techniques but again, do not give sufficiently frequent readings to act as a warning system if a landslide is imminent, although all of the techniques using survey methods do show long-term developments in instability if one measures points in the right location.

Often there are locations on large ancient landslides (of the sort that are *landscape features*) where buildings may remain sound for very long periods of time. A location I know well is the coastal landslide belt forming the Undercliff on the Isle of Wight. Large parts of the landslide system there are made up of rock blocks that are tens of metres across, and buildings on those suffer no damage at all even if the blocks are moving. Even some locations where there isn't such a raft may be satisfactory; for example the lighthouse at St Catherine's Point[12] moves seawards on a sub-horizontal slip surface typically around 100mm each winter, which may well be more than 20m since it was constructed in 1799, and although it wobbles slightly, it is in no danger of toppling. The damage comes at the boundaries of landslide units, and it should come as no surprise to discover that the walls of the lighthouse compound are riven by 'faults' and differential movements, nor that the access road is sometimes blocked and unpassable. The boundaries of landslide units may be fixed in place by the geology, for example being at the edge of a 'raft' of a landslipped rock stratum) or subject to change (as in landslides with a pattern affected by variable water inputs or toe erosion), but the areas unsuitable for development without stabilisation measures can be ascertained through careful mapping, or found by unfortunate experience when buildings are dislocated by movements so that good historic records are vital.

Where a community is threatened by the encroachment of a landslide toe, then instruments can be placed in the track of the landslide, and when their output reaches a critical level, a warning alarm can be sounded. They are not standard, off-the-shelf items, but tend to be designed and constructed for a specific site. One concept is upstanding strain-gauged piles in the path of the landslide that will be bent over if the landslide overwhelms them. The individual piles have to be strong enough not to be susceptible to vandalism. Another idea is to have tilt bars suspended over the landslide track *via* a cable and at a height where they are out of reach of animals and people. The bars have a mercury switch which activates when they are swung out of the vertical, but to avoid false alarms if the bars are swung by the wind, they have not only to reach a critical inclination, but also to have it maintained for long enough to rule out that they are simply swinging. Where I have seen such a system in the foothills of the Italian Dolomites they are combined with echo sounders as a secondary check that ground levels have risen, acting as a check on the possibility of false alarms from the tilt bars. I suppose that the tilt bars could be manufactured in such a way that they only actuate in one direction.

BARRIERS AND DIVERSION STRUCTURES

The catch fence is a form of barrier, but a catch fence designed to trap individual rocks would be overwhelmed by a debris flow and, in any case, requires maintenance to prevent a build-up of rocks if there are frequent rock fall events. Indeed, any structure that traps landslide debris will need periodic clearing, although a structure or earthwork that *diverts* landslides may or may not require clearing. A deflection structure in reinforced concrete upslope of a building in an avalanche-prone mountain location is one such diversion structure. In plan it will be arranged like the prow of a ship. Diversion structures can also be built in the form of embankments. Barriers built to capture landslides may take the form of compounds where the embankment surrounds an area into which winter debris flows can be captured, or across gullies to make a storage area akin to an impounding dam. If the expected landslides contain much water (and usually this sort of arrangement is built in response to frequent landslides at a particular location) then some form of water outlet is usually required.

Another kind of diversion structure seen in mountainous areas where landslides follow the same route is a shed structure built from reinforced concrete and designed to deflect landslides over a section of the transport infrastructure by having an armoured roof with a slope to pass debris. I have seen these structures equipped with old tyres to deaden the impact of rocks on the roof of the shed. The shed structure may be founded on piles drilled into bedrock or anchored into a hillside. (If you don't want them to work, found them on strip footings in landslide debris!) Inevitably, the structures are expensive and only provide protection where they are built, and any change in the pattern of landsliding may render them useless. Examples of processes that may change the pattern of landslides include earthquakes and hurricanes, or other devastating weather events.

Check dams are concrete and masonry structures built across stream valleys not to capture debris being washed down the stream channel but to prevent lowering of the thalweg, which would destabilise the stream valley slopes and add to the sediment burden in the stream. They look like capture structures that have been filled up, and thus to the untrained eye look in need of clearance.

BUILDING CODES

Building codes and planning regulations can be helpful in minimising risks to communities by preventing inappropriate developments, but the rules must be applied strictly and transparently. Many lay people simply cannot see the point of many regulations (even when they are sensible) and regard them as simple bureaucratic obstacles to their wishes. The same applies to developers, who are sometimes assisted by the oddities in the planning process in eventually producing buildings that are at risk. This is also known from the perspective of flood risk and developments on flood plains, and will continue to be a source of work for engineers rectifying problems into the future.

Sometimes there are obvious inconsistencies in the mapping of hazard, such as the boundary locations for an area of mapped flood hazard near where I live differing in height by 2 to 3m, and sometimes the hazard changes through time due to off-site developments, such as the introduction of flood control works or abandonment of them for financial or ecological reasons. If climatic changes with an associated increase in precipitation occur, then they would invalidate outdated stability assessments and hazard assessments on which planning regulations might have been based.

INSURANCES AND GOVERNMENTAL INTERVENTION

Property insurances on a single property can only rarely cover the costs of investigations, remedial works design and construction and building repairs along with the costs of alternative accommodation. It usually takes the insured values of two or more properties to undertake a complete solution, but reaching agreement causes delays, and arguments about which insurer pays what can lead to protracted wrangling and, in the end, possibly an expensive litigation. Money can be wasted if one owner takes the money from their insurer and runs, leaving the job unfinished. Up to a point, the insurer calls the shots, and some are more generous than others.

It is also a sad fact that insurers commonly weight their premiums over the whole of an area, defined by a postcode in the UK, or even exclude some risks. One example is landslide and subsidence risk in the PO38 area of the Isle of Wight, which contains not only the Ventnor Undercliff landslide complex but also a significant area with absolutely no landslide risk whatsoever in the next millennium or ten.

It may have been the case that the availability of governmental relief for flood victims has encouraged development in flood-susceptible areas, but the small footprint of landslide-affected areas has not provided the impetus to do the same for this particular hazard. In the UK, however, coastal defence grants are also possibly combined with slope stabilisation measures and planning regulations, which go towards improving the overall level of risk without necessarily encouraging unhelpful behaviour and preferably preventing it.

A very useful measure is to define a buffer zone along the crest of an erstwhile stabilised coastal slope in which clifftop development is forbidden. The same applies to some other slopes, including along the crest of infrastructure cuttings. I was party to the development of one such zone early in my career about a half-century ago. Nothing has failed yet. That encourages the foolhardy to question why such a zone was prescribed in the first place, and in a particular case I was able to report that the typical delayed failures of a slope in that geology typically occurred from 50 to 80 years after formation, which is what the buffer zone was devised to counter.

Other buffer zones in which re-development is forbidden may be the locations of concentrated differential movement in a huge landslide complex. Such an approach is tricky, because it does not allow for the development of new movement zones.

A WORD OF CAUTION ON INCLUDING COASTAL RETREAT CALCULATIONS IN A RISK ASSESSMENT

Coastal erosion is not a simple, regular, process, and a figure like 'a half metre a year' does not mean a half metre *every* year, no more, no less. The process is often a matter of a big 'bite' when a landslide happens, followed by a period in which the rear scarp of a landslide degrades to a quasi-stable slope with then a period when little or no retreat of the slope crest occurs. The toe, however, is probably catching up, and the slope is being readied for the next slide. Someone living at the cliff top may not be able to see the erosion of the toe.

In one case I looked at, a building surveyor had estimated rates of cliff top retreat from old Ordnance Survey maps, and on the basis of the calculated retreat rate had answered a question relating to the lifespan of a particular cliff top property as 'about a century'. In practice, it was far less, as the calculation had not only ignored the cyclicity of the landsliding and coastal retreat behaviour, but was also based on the assumption that the geology was uniform and so was the erosion at the foot of the slope. In practice, erosion at the toe of the slope had intercepted a plunging syncline structure along which a bigger-than-usual landslide had occurred, removing most of the buffer that had been assumed good for a century in little more than a single event taking only a few years at most.[13] Understandably, the house owners were not greatly impressed by the professional advice they had received. The past may well be a guide to the future, but sometimes it is not a very good guide.

NEARLY FINAL THOUGHTS

The successful stabilisation project naturally relies on a correct assessment of what is going wrong or what has happened, together with what might happen in the future, so that all realistic possibilities might be addressed. This assessment of what might happen in the future is equally, or some might say particularly important where the strategy is to do nothing and 'live with it'. The reason why this is an important issue is that many forms of instability are not the end of the matter, but they occur round the periphery of a developing and much larger instability. This can be seen as an example if you view some of the videos of rock slides on the Internet[14] where the failure is preceded by minor falls from below and both sides, and the development of tension cracks at the head of the slide. The corresponding features in slides in soils may include dislocated transport infrastructure, fences, pipelines, and buildings at the margin of the slide, and take place over many years, thus lacking the immediacy of the videos but lulling people into a sense of complacency.

Apart from a couple of cases that I looked at in Hong Kong, probably the most glaring case was at a dam site where construction had been halted due to a dispute – over a landslide, actually, but a relatively moderate size one perched above the eventual crest elevation of the dam. That landslide represented the head of a much bigger slide that had been mapped by the project geologist, and which was actually the smallest of a series of ancient landslides on that river bank, caused when the river had cut down through an ancient landslide dam and had been diverted into this river bank. Temporary roads had been cut into the slope, and the original morphology had been obscured. The excavations for the dam footprint would have completed the unloading of the toe begun with an excavation for a power station. As the documentation came into my possession it was clear that a set of cut slope failures neatly matched the outline drawn on the geologist's map. This was followed by a collapse in a particular location in the powerhouse cut, also exactly on the edge of the mapped slide.

It was a site that had its fair share of other issues, including surface monuments that appeared to have moved upstream and therefore sideways on the slope. They had been surveyed from the other side of the valley using a total station, and what seems to have happened is that the reference point on that side of the valley had moved, or the instrument wasn't zeroed properly. It all went to convince me that monitoring readings need to be assessed critically after they are taken, and those that show movement in an unexpected direction should be repeated to see if they are true. Like in so many cases I never found out what happened next as the dispute was resolved with the contractor leaving the scene with a hefty payoff.

As if that wasn't enough, I was called in to look at a disastrous slide where an embankment had been built over a soft clay foundation. Inclinometers had been installed, and were read diligently by an industrial placement student working for the contractor. Both contractor and resident engineer shared offices in an abandoned farmhouse, and the student had asked the resident engineer what she should do with the readings. He told her to put them on his laptop, which she did, and followed with every successive set right up to the failure, which, had he looked at them, would have told him very clearly that something was moving.

We should also beware of the fallacious 'not within living memory' argument (or the 'worst since records began' or 'the worst since 1910', *etc.*). Cherry-picking dates is a particularly grave misuse of data. But 'living memory' is a fickle thing because no-one really knows how long it is. The most egregious example was one I came across early in my career. A heavily armoured breakwater had been constructed, but it was damaged in a storm.[15] "*Will it be OK if it was just reconstructed to the original design?*" the team I was part of were asked, "*After all, it was the worst storm in living memory*".

It clearly wasn't anything like the worst storm since records began, nor the worst in the preceding decade, so the question was countered with: "*How do you know it was the worst storm, and therefore exceptional?*"

"*It breached the breakwater, didn't it?*"

The breakwater had only been finished for about a year, and allowing for the construction period, perhaps that meant that 'living memory' was as little as three years in that part of the world.

The 'living memory' fallacy is also related to the 'It's never happened before' fallacy, which some people take to mean 'So therefore it can't happen in the future'.

Repairing the breakwater to the way it was implies that the event or worse will never happen again; the 'never before' fallacy means that it can't happen at all. Ever. And that's just plain wrong.

Risk may well be quantified using proper assessments of what might happen, but the population in any affected region operates more through emotion than through some mathematical operation in which probabilities and consequences are assessed objectively by experts. Land and property values can be influenced by such emotional assessments, but equally, some other intangible considerations like having a sea view may make even some rather precarious sites much sought after. My deepest sympathies lie with those folk of a nervous disposition whose sleep or health is adversely affected by the fear of landslides. Blight extends to people's lives as well as to property values.

A COMPLETE ABSENCE OF NEED

There is, of course, not the slightest need for stabilisation in respect of a slope, or suite of slopes created during a construction project and which don't fail. However, geotechnical factors of safety are often quite small in practice, and relatively minor issues may cause failures that do need to be rectified at some stage, whether during construction, or afterwards, even decades later. Indeed, there is a valid argument that the absence of failures of any kind probably means that the design process was unduly conservative, especially where the costs of inconvenience and rectification are smaller than the additional costs of construction.

Engineers also often forget that the small geotechnical Factors of Safety in some design codes only work if they are used in conjunction with conservative estimates of the likely groundwater pressures and equally conservative estimates of soil shear strength parameters, as, for example, is recommended in Eurocode 7. When using continuum methods, being conservative in the use of any factor or factors may have consequences in the accuracy of any result. Beware! One merely hopes that one's failures are not disastrous and are easily rectified. Sadly, in the course of litigation, the standard that engineers are often judged by is that of absolute perfection.

NOTES

1 Lee, E. M. & Jones, D. K. C. (2004) *Landslide Risk Assessment*. Thomas Telford. There is an enlarged and improved second edition and even a third edition in the pipeline as I write.
2 An Internet search on 'ghost towns and abandoned villages in Italy' threw up multiple other cases.

3 Pugh, R. S., Gabriel, K. R., Jenkins, W. G. & Bromhead, E. N. (1991) Realignment of priory corner, Portland. In E. Bromhead, N. Dixon, & M-L. Ibsen (Editors) *Proceedings 8th International Symposium on Landslides, Cardiff, June 2000*. Thomas Telford, 1245–1252.
4 The Geobrugg company website has wonderful videos of tests of their rockfall barriers.
5 Notorious examples of slope instability damaging dams include the Malpasset Dam in southern France in 1957 and the St Francis Dam in California in 1928. The landslide at the Vaiont reservoir in October 1966 caused over 2000 fatalities, and although the dam survived, the wave caused damage and fatalities 200m above the dam crest.
6 As it so often does. In this case my comment that there was "*absolutely nothing happening*" had become transmogrified into: "*Ventnor landslide could kill: Expert says*". It was in The Times, of all places.
7 Ordinarily, the water-bearing utilities are potable water supply and sewerage. Where sanitation utilises seawater, there is a third pipe network, and another set of service reservoirs to supplement those for potable water.
8 This landslide crossed the Po Shan Road and shunted some buildings belonging to a boatyard into the sea. Two hikers had camped illicitly overnight in the building and were the only fatalities. The effects of a landslide in terms of fatalities are therefore sometimes dependent on the accident of timing, as in this case a daytime occurrence would have caused more fatalities amongst the workforce.
9 Perhaps my choice is predicated by what I see when visiting my favourite rock slopes in the Lune Gorge on the A685 road where it is parallel to the M6 motorway in NW England.
10 Bizarrely, there are some wooden chimneys that have survived since the Middle Ages. They were made of wattle and daub. Presumably, many didn't survive and went up in flames when the clay daub cracked and dropped off.
11 On a field visit to the Italian Dolomites, our guide pointed out that many screes and debris accumulations were not natural features, but were the result of gunfire during the First World War, or they were the spoil from the tunnelling operations that took place at the time. The result was a whole set of things that looked like landslides but which weren't.
12 See, for example, the paper by Hutchinson, J. N., Bromhead, E. N. & Chandler, M. P. (2002) Landslide movements affecting the lighthouse at Saint Catherine's Point, Isle of Wight. *Conference on Instability: Planning & Management, Thomas Telford, May 2002*, 291–298. In order to establish a network of fixed points to allow rapid survey using GPS, the UK national mapping agency (the Ordnance Survey) in conjunction with the lighthouse authority (Trinity House) put fixed stations on lighthouses around the coast of England and Wales. The one at St Catherine's gave a movement record in the winter of 2000–2001 that confirmed the results of a geodetic survey some years before, as the lighthouse proved to be anything but a fixed point.
13 I was asked a similar question once, by an elderly couple whose bungalow was close to the edge of a retreating cliff. They were both ill, with a medical estimate of a short remaining life expectancy. They hoped that the house had a marginally longer lifespan than they did. The next time I visited the site, the bungalow was gone. My memory of the encounter is marked by a great sadness.

I was also asked for an estimate of the expectancy of a nearby radar station. I had forgotten my reply, but one day received an excited phone call from the enquirer. I'd estimated something like 5 to 7 years, and the station began to collapse exactly 7 years from the date on my letter. Since then I have been doing the football pools or the lottery, but I think I probably exhausted all my latent clairvoyant abilities (and my luck) in that one time.

14 Usefully captured by David Petley in the AGU *Landslide Blog* or on YouTube.

15 It wasn't a landslide, but it was the failure of a rockfill embankment. A *badly engineered* rockfill embankment, as it happens.

Chapter 9

Remedying landslides

OVERVIEW

On the grounds that it is sometimes (as in an earlier chapter) said that the three worst things to do to a landslide are to place a load at its head, to excavate away at its toe, or to raise the groundwater levels, then probably the best and most obvious ways to stabilise a landslide are to unload its head, build a berm at its toe, or provide drainage, or to do two of them, or all three. Those three worst actions are not the only things that can destabilise a dormant landslide. For example, to do anything that reduces the strength of the soil along the sliding surface, or remove some sort of obstacle that supports the slope, can have a similar destabilising effect, so improving the strength of the soil; or providing structural support are also options in addition to the three most obvious solutions. In a paper I wrote for the ICE a number of years ago,[1] I pointed out that stabilisation was only one of a number of responses that could reduce risk from a landslide and the ones that didn't involve stabilisation were described in the previous chapter. I have had several chapters in books on the subject of landslide stabilisation,[2] but always there is the issue of space, and every possible solution can never be described.

NO 1 OF THE BIG THREE: UNLOAD THE HEAD

Taking first the idea of excavating the head of the landslide to reduce the driving forces, the downside of this approach is the potential to enlarge the head scar and thus destabilise it. Even just leaving an existing steep head scarp (scar or scarp?) is never a good idea, because even minor collapses will load the actual head of the landslide and then possibly undo the good that excavation did.

I had several experiences of this when visiting coastal slopes at various locations in southern England. In one particular case, the original head scarp had been left untreated because to batter the slope would have reduced the area of greensward at the clifftop. The area was considered to be a

DOI: 10.1201/9781003428169-9

considerable local amenity and could not be touched. Personally, I always felt that the steep slopes (you could call them a cliff) represented a hazard to people and dogs if they fell off it, and if it had been left to me I think I would have done something about it. The scheme worked adequately for some decades until some rather minor collapses of the head scarp set off a chain of events that resulted in a complete breakup of the stabilisation works, a process that was repeated at numerous locations along a couple of kilometres of 'stabilised' coastline.

In another location where the coastal landslides on an unprotected coast were eating back into the toe of a hill, the landslides were not actually stabilised, but the rear scarps were flattened with the surplus material simply dozed over the landslides. This has removed the hazard to people of falling over the head scarp, and two decades later very little coastal regression appears to have happened: the slides were not very active.

One issue relating to removing the head of a landslide and battering back the head scarp is that the stabilisation works then have a bigger footprint than the landslide itself. This issue can complicate matters, for example, with an infrastructure cutting in an urban area where roads, buildings or even their gardens run right up to the crest of the slope. It may then be necessary to reinstate the excavated-out crest using lightweight fill. Digging out the head of a slide and restoring ground levels with lightweight fill is an excellent way of stalling a landslide in course of development which has manifested itself with a pattern of cracking, and it has the additional benefit of providing a source of fill for a toe berm or other regrading if there is space for it.

A rather neat solution to a problem of the instability of an embankment on a weak foundation was to remove approximately the upper half of the embankment fill and to rebuild it using polystyrene blocks. The embankment needed to be at its original height because it was the approach to a bridge over a motorway under construction. The unit weight of the polystyrene blocks was about half the unit weight of the embankment fill, and by analogy with a bearing capacity problem, removing the half height made the Factor of Safety increase to approximately 2, and then putting back half of the difference using 1 tonne/m^3 polystyrene instead of 2 tonne/m^3 soil brought the Factor of Safety F down to a still-reasonable 1.5 or thereabouts. While this was the rationale behind the design, the detail was confirmed by a rather more precise formal analysis. Keen golfers in the area therefore retained access to their course, even though a motorway had been built between the site entrance and the clubhouse.

A variety of lightweight fills can be used. Polystyrene has the disadvantage that it is flammable or may be dissolved by some common chemicals, a problem which is particularly pronounced with low-density foam blocks. Lightweight polystyrene blocks must therefore be adequately covered. Compressed bales of end-of-life vehicle tyres can also be used to restore ground levels at the head of a landslide that has been excavated. These typically have the same or similar density to polystyrene but less tendency to float if inundated, because of their porosity. Another approach is to import foamed and sintered clay aggregate pellets. When loose, they can be lighter even than polystyrene, but they also float. These low densities are not achieved if the contractor compacts the lightweight aggregate!

An excavation at the top of a slope that is to be backfilled with a lightweight fill can be made lighter still if large voids are constructed in it. If the ground is unlikely to be heavily loaded in future, large voids can be created using galvanised corrugated steel culvert units (or other shapes) or other constructed underground spaces. In some cases, these underground spaces are useful for domestic or industrial purposes, although in such cases it is advisable for them to be properly ventilated to prevent the accumulation of heavy or health-damaging gasses, in particular carbon dioxide[3] or carbon monoxide from vehicle exhausts.

Moreover, I have seen, to my complete astonishment, wood chippings used as lightweight fill. They do not seem to me to have a long service life.

In the past, the practice on Britain's railways when dealing with a slide in a cut slope was often to dig out the slipped material,[4] relying implicitly on the undrained strength of the ground that had not failed to support the sides of the excavation (implicitly because this was experience, as the requisite theory had not been developed then). The excavation would be backfilled with a variety of materials, most commonly 'spent' ballast, or the aggregate that supports the track but which had become contaminated with soil or the various things that dripped from trains and carriages, including from the lavatories. There are several advantages to this procedure: not only is it a reuse of granular materials that would otherwise be dumped, but any slip surfaces are broken up and any weak strata removed. Nowadays, spent ballast counts as contaminated, and new materials must be used, but that includes rockfill, which can be tailored to the situation, including being reprofiled to a different line than the original slope, installed with no fines to make it permeable, and placed with geofabric filters to prevent ingress of fines from the unslipped ground underneath and beside the rockfill.

There is also some considerable benefit in reusing the excavated materials on site, because once again, slip surfaces that are usually associated with weak strata are broken up and mixed in with the replaced fill. Vehicle movements are reduced, and the costs of importing materials and disposing of waste are eliminated. Clay materials can become a little drier and incorporate air when recompacted and so exhibit a better undrained strength, at least until they wet up again, but the wetting can be delayed or prevented by associated surface drainage measures or by incorporating drainage layers in the replaced infill. On the other hand, importing rockfill provides what is usually a substantial increase of strength. You have to be careful not to create preferential alignments along which a new slide can develop. There is also the prospect of using geogrid reinforcement, installed layer by layer as the backfill is placed and compacted, for added benefit. Multiple layers of plastic geogrid laid on top of each other may constitute that preferential 'slippy' alignment, and geogrid layers should preferably be interspersed with soil, except at obvious overlapping or splicing locations.

I am sometimes asked whether I know how long a temporary works excavation can stay open, or alternatively, if I know how to calculate it. I think the problem is rather more complicated than working out the equilibration of construction-induced suctions in a cut slope (which is hard enough), but a steep-sided excavation contains the added difficulty of joints and fissures opening to perhaps allow some preliminary toppling movements, and the relatively small size of the whole making it very susceptible to individual weather events. My preferences would be to batter back the sides of any excavation, to support the sides (if steep) probably with something like soil nails or anchors, or to use a long-reach excavator and forbid man entry.[5] In turn, forbidding man-entry into the excavation might rule out the use of geogrid reinforcement that usually requires laying manually. If the excavation slope is designed on the basis of sensible measurements of the undrained shear strength and in addition receives some support from a three-dimensional plan shape then it is probably more stable than a long, linear, excavation, but caution is preferable to an accident.

The analysis of this approach to slope stabilisation is a question of modifying the data set used for analysing the failure to include the revised internal structure, the revised groundwater conditions and, if different to the original, the revised ground profile and running the whole thing through the computer again, repeated with suitable modifications until a satisfactory result is obtained. My personal preference is for the designer to at least check that the temporary works case(s) are, in principle, safe, which means analysing the slope at the various stages in excavation and then backfilling. If there is a critical case, then the contractor should know by being told, and the works need to be supervised by someone who has been properly briefed and is aware of any constraints on what is done.

I often look at temporary works excavation for, say, a retaining wall and wonder who designed it – the excavation – as surely that slope, that hillside, needs the wall, and at the moment without it, we are rather reliant on how hard the soil can suck and, in effect, how long it can hold its breath.

NO 2 OF THE BIG THREE: LOAD THE TOE WITH A BERM

When it comes to stabilising fills at the toe of the slope, lightweight materials are probably the last thing that anyone contemplates because an important factor in a toe berm is weight. Toe loads, toe weights or toe berms or whatever else they are called almost always require land take outside the footprint of the landslide. Such land may not be readily available, say, with a landslide in an infrastructure cutting. Various solutions to this problem include raising one carriageway and running it over the top of the berm. Another solution on a rural railway was to close one line and develop a system of two-way working. In a third example, one lane of a dual carriageway road was semi-permanently closed to traffic, and in another the hard shoulder of a motorway was closed temporarily.

With coastal landslides, the toe berm solution requires projection out over the shore. Changes to the alignment of a coastline may adversely affect the coastal hydraulics, causing beach accumulation on the weather side and beach denudation on the lee and, in any case, probably reducing the amenity value of any beach to bathers and other users. Where there is a river at the toe of the slope it may be realigned and erosion protection incorporated in the new riverbank.

In terms of the analysis of a stabilising berm built out over flat ground beyond the toe of a landslide, my method is to increase the size of the berm, both its crest elevation and its crest width, until I get the sort of Factor of Safety improvement that I desire relative to the equilibrium of the particular landslide. As well as projecting the toe of the landslide through the berm, I look at different breakout angles, a simple push where the berm slides along its base and slip surfaces based on the real landslide but which override the berm and 'underride' it (the sliding on its base being a special case of that). In all of this analysis, I model the distal batter of the berm as steep as I can, because then the batter gives me that little extra, and the scheme is not dependent on what the slope of the batter is, and so I can decide the batter based on a slip circle type of routine analysis (if necessary) where the berm is built of local soil. I will have therefore separated the choice of batter from the selection of berm size. Of course, once the overall size is selected, there is room for 'value engineering' the system down to an optimum size.

Embankment side slopes are sometimes built to a larger-than-desired final size so that the soil is well-compacted, and then the excess is trimmed off to the desired shape. In critical cases, this sequence may threaten marginal stability, and the appropriate temporary works cases need to be checked to ensure that the procedure can be completed safely. The side slopes may then need to be constructed with stronger material, such as rockfill.

If the available footprint for a berm is rather small, it may need to have a high crest, but the berm must be wide enough to allow proper compaction by machine. A temporary works case that needs some attention is where the berm fill is placed over benches cut into the toe of the landslide. This case represents excavation at the toe, and that is one of the three main destabilising factors, which must be done with care. Excavating the benches in short lengths is possible, and may prevent wholesale movement of the landslide if done judiciously, but inevitably some creep movement is likely. It is therefore better to provide some other improvements to overall stability such as head unloading or drainage before benching in at the toe. I know that I am repeating myself, but it is an important consideration.

Because a berm functions through its weight, it is desirable to use dense fills. Rockfill is a good material, as appropriate densities are achieved with quarry-run materials that have a range of particle sizes, and it can be used with a steep batter, and that minimises land take. When compacted properly, rockfill has a very high angle of shearing resistance, but when used across the toe of a landslide, the magnitude of that resistance gives problems in the analysis itself as projecting (say) an upwards-directed slip surface rising up through such a material gives a misleading and unconservative value for the increase in Factor of Safety. Even worse, where wave armour with large blocks or concrete armour units is used in a coastal location, this apparently high angle may be the result of only a few contacts between blocks and cannot be trusted to retain the slope. It is essential to test the under-riding and over-riding toe positions, and also to test whether the computed 'through-the-berm' toe breakout position Factor of Safety is genuine or just one of the oddities of limit-equilibrium methods.

Marine or riverine aggregates are often rounded and do not give the same degree of interlock needed to reach the highest angles of shearing resistance, but they may be satisfactory as backfill if retained by a gabion wall or other support structure.

As the limit-equilibrium basis of any method of slices does not deal with strains, the computed improvement in stability is that which can be obtained when sufficient deformations have been built up to provide the necessary (and calculated) resistance. This may mean that deformations continue after construction, or that rather bigger resistances have to be provided to control those deformations.

NO 3 OF THE BIG THREE: DRAINAGE

Drainage is almost always a good option, but it always carries the risk that the drainage system is not maintained or has been constructed in a way that is not maintainable, and then it is liable to blockage. Blockages to drainage have a variety of causes. One of them is the precipitation of minerals from the water that flows through the drainage system. Water is sometimes colloquially described as 'hard' if it contains calcium carbonate, and 'soft' if it does not. People who live in areas where hard water is delivered *via* the public water supply will know that chalky deposits are readily precipitated in kettles and in washing machines; those deposits don't need to be boiled to leave their traces. Gravel backfill in drains is particularly susceptible to clogging with calcium carbonate precipitate, which produces something rather akin to concrete when the precipitate has filled the voids. Iron compounds are also readily precipitated, sometimes assisted by various forms of bacteriological action. Iron-rich waters commonly emanate from old mine workings, and the water and its deposits have a rusty red or ochreous colour.

Other blockages can occur from fine particles washed into the drain. This is particularly acute with gravel-filled trench drains open at the surface, but can be a problem in drainage galleries where fines are washed out of the parent rock mass by the inflow of water. Where feeder drains carry such fines into a drainage gallery and the water velocity drops, the fines may sediment and then block outlets. Another form of outlet blockage comes if the temperature drops past freezing point and the outlets block with ice.

A problem that arises with deep drainage in a large landslide system is the extraction of groundwater causing settlements that are detected at the surface. Hence, if there are buildings or infrastructure constructed on the landslide,[6] consideration needs to be given to whether or not this settlement is likely to be experienced and, if so, whether it can be tolerated in the short term in order to capitalise on the long term benefits of stability.

Drainage solutions for landslide stabilisation may be thought of in two categories: firstly, removing water from the landslide or secondarily, preventing water entering the landslide in the first place. Even having made this subdivision into categories, a further subdivision into surface water or shallow drainage and deep drainage may be called for, although shallow drainage is often in the category of preventing ingress and deep drainage to remove water that is already there. Moreover, some drainage only needs a short life span because it is there to remove water generated in soil under loading (that is, when the loading is *undrained*), whereas some drainage has to work for the lifetime of the remedial scheme, which in many cases is effectively forever, and drainage can never do that without a maintenance regime.

Where drainage is installed to get rid of undrained loading induced water pressures, we could add that removing water from where water pressures are generated is often simpler than trying to catch them in other places.

Preventing the entry of water is only possible to a limited degree. I'm not greatly in favour of a sprayed concrete blanket over the top of a slope, but I am in favour of plastic sheeting over small landslides during the rainy season to prevent infiltration. Such sheeting is used in agriculture to cover haystacks and so is available in large, robust, sheets which can be taped together if necessary, and held in position with stakes and ropes, or weighted down with sandbags. It is, however, only a temporary measure. Sheeting is particularly useful if laid before snowfall, as all the water from snow melt then runs off. Snow melt does sometimes have particularly bad infiltration characteristics, as the remaining snow cover inhibits runoff, and so the surface is wetted continually and for longer than in the case of rainfall.

Vegetation is also useful in preventing infiltration, as much rain can be held in the canopy of trees, where it evaporates back into the atmosphere without even touching the ground. Where the rainy season is also the cold, windy, season in places such as the UK, a deciduous tree cover is less effective at rainfall interception than evergreen shrubs, but even so, the cover has limited capacity and is overwhelmed in heavy downpours, which is why the vegetation cover is always useful, but not to a calculable or, in most cases, significant, extent.

Shallow ditches are also useful, both to capture rainfall and to carry it away. In combination with land drains they can have beneficial effects in open fields, around sports pitches and in grassland.

In terms of estimating the effects of drainage on stability, it needs to be remembered that the drawdown in the water levels is variable in the space between drains and at a maximum in the drains themselves. It is therefore a three-dimensional problem that requires an average reduction in piezometric levels to be fed into your stability analysis datafile. The distribution of drainage effects in the plane of the section (for 2D analysis) may also be an important factor in particularly discovering the effects on some potential failure surfaces rather than others.

Drains in a landslide may have effects on the ecosystem of the landslide. In at least one case the drains so dried out the surface that the site was abandoned by its newts (a protected species), but they were replaced by butterfly larvae in their thousands from a nearby site where the butterflies had, until then, been thought to be in danger of extinction.

Way back in railway construction, slips were sometimes stabilised by digging trenches in them and backfilling the trenches with various materials that not only gave them a drainage function but also reinforced the slope if they penetrated through the basal slip surface. They are therefore 'counterforts' or 'counterfort drains'. Nowadays, the drainage material would be encapsulated in a geofabric filter. There are mixed views on whether or not to provide a pipe at the bottom of the drain: I have seen butt-jointed clay pipes used, but they require man entry into the trench, which is something that I think should be avoided on safety grounds. In the particular case that I recollect, the base of the trench was sealed with a little concrete. I have also seen trenches where an impermeable membrane has been laid at the base, and in other cases, the base of the trench was simply left bare. The modern substitute for the clay pipes is perforated plastic pipe supplied on drums and therefore not needing splicing, or as lengths of tubing with threaded ends and couplers.

The use of perforated or otherwise permeable pipework in gravel-filled trenches is probably acceptable where the water inputs are simply derived from the immediate locality of the drain. However, when they are used to also carry the water collected over a large area, for example, where the water is collected over a long length of interceptor drain along the crest of a cutting, the capacity of the pipe may be overwhelmed, and as in the case of a blockage, the water may back up and move out from the pipe. If flow rates are then sufficient, the gravel backfill of the drain may be washed out, or the surrounding ground may be destabilised. The problem is particularly acute in railways, where even a small quantity of redeposited material may be enough to cause a derailment.

Interceptor drains at the head of a cutting are often a source of trouble if they cannot discharge freely and safely, giving rise to localised failures where water spills out due to various forms of blockage, low points or even animals burrowing. Multiple exit points from a crest interceptor drain are therefore useful in carrying water down a cut slope.

GROUND ANCHORS

Ground anchors can be a very effective solution for some types of landslide, because they are active, inserting restraining forces directly into a landslide and not requiring any landslide movement ground deformations to mobilise their resistance. They do, however, cause their own ground deformations locally, both when tensioned and when the ground consolidates under the tension loading. As a result, ground anchors may require retensioning several times after installation and initial stressing.

Basically, the technology is similar to that of post-tensioned concrete, although with the difference that a hole needs to be drilled in which to insert the anchor tendon, and there is no 'other side' to create the anchorage. Instead, the distal end of the tendon must be secured in the borehole. In soils, the borehole may be enlarged at depth (or 'under-reamed') in order to increase the pull-out resistance, but in rock it is probable that the drill hole will remain parallel sided. Commonly, the tendon is grouted in place, and to make a better connection between the grout and the end of the tendon, it is usual to strip off any plastic sheath that protects the steel wires from corrosion, degrease the individual strands and cast them in a capsule with epoxy resin. The cable is inserted into the drillhole along with a grout tube, and once in position, the capsule is grouted in place. Stressing the anchor cannot be done until the grout has hardened, and various additives may be used to accelerate the process. In some instances, a chemical grout can be considered.

At the ground surface the concentrated tension force must be spread out into the ground, normally done using a cast in situ reinforced concrete pad, but precast pads are sometimes used. The deformations upon tensioning can be expected to be more with precast pads than with cast in situ pads. Each pad is dimensioned so that it does not suffer a bearing capacity failure under load, and it is reinforced to carry the appropriate loads. Pads are normally constructed or placed in their final position with preformed holes through which drilling takes place, which ensures that the holes in the pad align with the holes in the ground, the concrete of the pad does not need to be drilled through, and the reinforcement to prevent 'bursting' is located in the right place.

Pads may be designed to take more than one anchor, although I am not enthusiastic about multi-anchor pads because stressing some anchors may destress others should the pad tilt or rotate when loaded.

Ground anchors can also be used in combination with some structures, for example piles and walls. Care must be taken to ensure that the anchorage at depth is underneath the slip surface, as otherwise the whole anchor force will be an internally balanced system that does not help in stabilisation of the whole mass, although local anchorages may still be useful in respect of smaller components such as the faces within slipped masses, retaining walls and other features. Another defect that may arise with an anchored wall is that movements of a landslide before all the anchors have been installed or stressed may use the wall as a lever and thus destroy the anchors that have been installed.

STRUCTURAL SOLUTIONS

Passive structural solutions to the landslide problem fall into several well-defined categories depending on whether their principal mode of action is tension, bending or shear. There does not seem to be any equivalent in compression, beyond a strutted excavation. Soil nails are basically tension piles; ordinary piles and embedded walls may operate in shear or bending, and shear keys operate in shear – as might be imagined from their name.

SOIL NAILS AND ROCK BOLTS

A soil nail can be thought of as an initially unstressed anchor, which has a latent retention capacity which can be mobilised if there are ground movements. It requires some means of restraining the soil or rock between the individual nails, which can be done with mesh and sprayed concrete or steel mesh and plastic sheeting.

A soil nail is typically a length of alloy steel bar inserted into a borehole and grouted in place. The bar has a threaded end, and a load-spreading plate of some sort at the surface is secured with a nut which provides some initial tension in the nail. Often the initial stress in a soil nail is only there to hold components in place rather than to stabilise the slope, as in the case of a tensioned anchor. The main loading in a passive system comes as a reaction once there is some movement.

Personally, I find the name 'nail' confusing, as for me, a nail is something you drive in with a hammer. I am also somewhat prejudiced against soil nails in view of the number of failed installations I have seen, including both soil and rock falling out between the nails and the failure surface being deeper than the nails so that they do nothing to keep a slope stable. Perhaps it is because a successful installation soon becomes invisible, and I have never been commissioned to go and see successful nailing, or perhaps, as I suspect, in some of the more successful cases the nails were redundant anyway. I certainly believe that they have their uses, especially in temporary support to slopes while some more elaborate structures or earthworks are constructed, or when the most likely mode of failure is shallow.

Rock bolts are also used in tunnel supports as well as in rock slopes, and I have included them here because of their general similarity to soil nails. There is again a question of desirable length, which is more easily determined in tunnel support than in a slope. Ease of installation is also a factor, and so is the speed with which they provide support. One method of providing grip at the distal end is to have a sawcut in the bar with a wedge so that when the bolt is driven into its drill hole against a wedge, the ends of the bar splay out like a fishtail. There are also systems with two-pack epoxy that is inserted in the hole in sachets that are chopped up and mixed with a chisel end to the bar or expanding shoes of the Rawlbolt type. If the speed with which support is developed is less important than cost, bolts can be grouted with cement, although to minimise shrinkage a small amount of expanding agent can be helpful.

EMBEDDED WALLS AND PILES

In order to be used to stabilise a landslide, the embedded part of a wall must cut through the slip surface. Then, if the wall is intended to operate as a cantilever, it must have sufficient embedment to be encastré, which becomes progressively easier the stronger the soil beneath the slip surface, although this accordingly makes the installation of the wall more difficult.

Steel sheet piles are sometimes used with small landslides, where they are often effective, for example in railway and highway cuttings where the stabilisation works have to be constructed within the original footprint of the slide. The flexibility of steel sheet piles is less if Z profile units are used, as U piles have the interlocks on the neutral axis in bending, which reduces their bending stiffness. Walings are useful, not only to provide anchorages for passive ties or ground anchors, but also to prevent bulging in plan, which stretches the sheet pile wall, and this reduces the section modulus. Corrosion is always a problem in an intertidal zone or near the coast.

Most embedded walls are made from cast in situ reinforced concrete. Where the equipment is available, a diaphragm walling technique is useful. This usually entails a crane-mounted clamshell grab to do the excavation at the base of a trench which is filled with a bentonite suspension to keep the sides open. Concrete is placed *via* a tremie pipe to displace the bentonite, which is saved for reuse, and a reinforcement cage is inserted into the wet concrete (or into the bentonite before the concrete is tremied in). The walls do not need to be straight in plan overall, although the clamshell grab may mean that they have straight sections, and the walls may form archline shapes in plan, or have projections downslope (buttresses) or upslope (counterforts). There projections increase the length over which shear is to be carried and increase the bending resistance and stiffness.

You should note that it is wrong to design a landslide-stabilisation wall using conventional earth pressure theory, as the loads on it do not come from steep soil wedges but from the landslide as a whole. Loading on the head of a landslide may significantly increase the loads on an embedded wall even if the head and the wall are a long distance apart. Where a retaining wall does derive its loading or resistance from conventional earth pressure the modification to earth pressure coefficients if the ground slopes (both on active and passive sides) may be a critical factor.

Bored piles can be an effective means to stabilise comparatively shallow landslides. They may be installed widely spaced in one or more rows, or they may be installed to form a wall by being closely spaced, contiguous, or cut slightly into each other to form a secant piled wall. Reinforcement by means of cages of bars that are the norm in foundation piles may be used, but rolled steel H sections or tubes are easier to install, as they can be picked up and dropped into a wet concrete filled pile shaft.

A big advantage of piles and other support systems that penetrate through the slip surface of a landslide is that they are sufficiently rigid. Their resistance is mobilised with comparatively small ground deformations, something that is not obvious when their design is done with the aid of limit equilibrium methods and is far more evident with continuum methods.

Where a row of piles is the main stabilisation method – and this applies whether they are spaced out or form a continuous wall – they have most movement at their heads, which is not addressed by a straight capping beam. Deformations are better controlled by making the pile layout have what are effectively the counterforts or buttresses that can be installed with diaphragm walling, or indeed, layouts that are not simply straight in plan, but have arches, sawtooth or castellated 'square-wave' planform. A capping beam also provides a useful anchorage point for ties or stressed anchors that will control deformations.

One design approach is to calculate the individual pile collapse mechanisms in shear, bending, or rotation, then to take the least resistance as a reaction force to feed back into the stability analysis, eventually giving the number of piles needed. It has the advantage that the reinforcement only needs to be provided for that worst mechanism. Alternatively, the forces on the wall or piles come out of the stability analysis, and then the structural elements are designed to suit.

Shear keys and buttresses may be reinforced or unreinforced concrete or masonry and take the form of embedded walls that are aligned in the direction of movement of the landslides, most commonly in the region of its toe. They have generic similarities with structural foundations around which a landslide may slide, but they are sufficient of them to prevent movement of the slide through them.

CRIB AND GABION WALLS

I have reservations about the use of crib or gabion walls in landslide stabilisation work, and prefer them to be used to support the face of an earthwork that in turn uses its weight to stabilise a landslide and not as a stabilisation measure on their own. Gabions rely to a large extent on the longevity of the wire basket, although if that corrodes the infill becomes a dry stone wall, and it then depends on the efficacy with which they were packed. Plastic baskets may suffer from exposure to sunlight, and, in any case, are likely to melt or burn if exposed to fire. In third-world rural areas the populace may find a gabion structure a useful source of already quarried building stone, and since gabion walls are gravity structures, it will be obvious that removing their weight is not helpful. Gabion structures do accommodate some movement, and are rather obliging sometimes in that they accommodate a *lot* of movement.

Crib walls can be made from a variety of materials, including timber, steel sections, and precast concrete units. They act compositely with granular backfill to make a gravity structure. The life span of wood in contact with the ground is, at best, a few decades, and in some cases, much less. I once saw a timber crib wall supporting a house and considered that it must have been designed by someone who anticipated being nowhere around when it came to needing repair or replacement, besides which it was vulnerable to vandalism by fire. Galvanised steel units probably have a limited life too, but somewhat longer than wood, unless in a coastal environment where they are alternately submerged or splashed and then dried, in which case their life is likely to be equally short.

GEOGRID AND SOIL REINFORCEMENT

The invention of geogrids and the idea of soil reinforcement has had an impact on slope stability practice in that it has allowed steeper sideslopes in fills to be constructed when the geogrid is used as 'face reinforcement'. If a fill is constructed on sidelong ground, its face and the underlying ground may fail together, and some geogrid reinforcement may prevent that if the stability problem is not that of the whole fill. Expect displacements, however, as the reinforcement takes up load, which it may do at a particular point in construction with a 'kick' in inclinometer readings or the outputs of other monitoring equipment, topographic surveys, *etc.*

GROUTING AND STABILISED SOILS

Bizarre as it may seem, some success has been achieved in grouting up landslips. There was a period when this technique was used extensively in the UK rail network, but the technique has fallen out of use. In principle a cementitious grout is injected into the slip mass under pressure, and the grout finds its way through voids and fissures, eventually solidifying as a network of stronger material. Presumably the grout works its way along slip surfaces. There may well be an element of ion exchange in some materials that improves strength in situ akin to the effects of lime piles. For obvious reasons, the process must be undertaken gradually, or you risk blowing the landslide body apart. I have been aware of the technique for some decades, but have never seen it employed in practice, and I can see that it could have very immediate beneficial results in soils with large voids or porosity.

The ion-exchange and hardening properties of grouts are mirrored in the use of additives such as lime and cement to improve the properties of fills. Both of these materials can be aggressive to site workers and need to be used with great care. The term used for them is 'stabilised soil'. Lime is used to remove surplus water, but cement is used mainly for its increase in strength.[7] An effective application in landslide stabilisation might be to improve the properties of excavated materials reused as fill elsewhere or as part of a dig-out-and-replace scheme.

The old paper[8] in *Géotechnique* about 'baking' piles in a large clay landslide on the Black Sea coast with gas fires sourced from a nearby oilfield has echoes in the 19th-century method of coal fires in pits in railway cutting landslides. The only time I ever came across this was in a disused railway line, with a local builder excavating the brick-like material for use as bulk fill and as a base course for estate roads. Fortunately, the associated railway line had been closed.

ROCK SLOPES

Landslides in rock slopes that take place along such stratigraphic features as clay seams have most of the characteristics of landslides in soils, including the same methods of analysis and treatment, with the added complexity of the difficulty and therefore expense associated with investigating them and of construction of remedial works that involve excavation into the rock mass.

Other rock slopes have their stability determined by the pattern of discontinuities in the rock mass.

CONSTRUCTION SEQUENCING

The most obvious need for proper consideration as to construction sequencing comes from the case of the construction of a toe weighting berm to permanently stabilise a landslide that has come to rest temporarily. The contractor clears the toe of the landslide of vegetation and topsoil then proceeds to remodel it into a series of benches so that the toe berm can be placed and compacted properly. While doing this benching in, the landslide experiences toe unloading and moves in response. The answer was to take account of the need for a good practice like benching in but to realise that this would be a destabilising action that would need a corresponding positive action such as some unloading at the head before the other works could be started.

Another temporary works situation that could give rise to stability problems is the need for a piling platform to be constructed before heavy plant such as a piling rig can be positioned to start work. While the platform is necessary to prevent a piling rig toppling, it does represent a heavy loading on an unstable slope.

In principle, my advice is to at least design something that is constructable in safety, and if a contractor proposes an alternative procedure, check that the works continue to be safe at all stages.

ASSET MANAGEMENT AND MAINTENANCE

Earthworks, including cuttings, embankments, dams, and so on, are just as much 'assets' as are, for example, the stations, bridges and culverts on a railway line, and they may need inspection, maintenance and repair in just the same way as structures do. The recognition of this fact has led to infrastructure network operators (roads, railways, canals, pipelines, and power transmission) considering the problem of maintenance holistically. This has led to the idea of entering slopes and earthworks into asset registers and having a system of inspection in which they are expected to show long-term deterioration if nothing is done to maintain them, so that the need for maintenance ceases to be a surprise but, in contrast, is an expected and routine call on resources. This contrasts with the approach in which nothing is done and failures come as a surprise, and then are unexpected events to deal with on an emergency basis.

Infrastructure operators may well have a range of geologies to contend with, and a range of assets or various ages and types that have different needs, but generally, however, the process improves budgeting and cashflow.

In many respects, the Landslide Inventory of the Geotechnical Engineering Office (formerly the Geotechnical Control Office) of Hong Kong falls into a similar category, although there the slopes were not necessarily under the ownership of the GEO. Part of the intention of the inventory was to classify slopes in terms of their risk (likelihood of failure and consequence if they did) so that they could be prioritised for treatment. A side effect of the studies and inspections was also to make available a database of things that led to landslides in the territory. For example, it was found that once a rainstorm passed a particular intensity threshold, it was likely that multiple landslides would occur. This did not mean that for less intense rainstorms no landslides would occur, but it did mean that emergency services could potentially be put on alert. The inventory also revealed some oddities, such as the water from firefighting having the potential to provoke instability.

As far as I know, the inventory approach is relatively unsuccessful at indicating *where* major landslides might occur, as they tend to be statistical outliers. What this approach does show, however, is that the big events are normally preceded by much smaller events, such as cracking of retaining structures, small slippages, and so on, which can be correlated retrospectively with the development of the big event. Since the smaller precursor events occur on the periphery of the eventual large event, proper inspection and interpretation of a landslide inventory is important.

It eventually becomes obvious to the owner of a large number of assets of a similar kind that some form of management, inspection, repair, and maintenance regime is necessary in order to manage costs, but it is less obvious to the owner of only one example, especially one that is so robust as to go without maintenance for years or decades. An example that comes to mind is some form of retaining structure in a garden, or between two rows of houses. Matters are worsened if the structure is a shared asset, for example a retaining wall that is part-owned by two adjacent properties or one that is sufficiently old for its origins to be unknown. They are worse too if the asset belonged to a now-defunct organisation, with no traceable ownership, such as a mine waste dump, settling lagoon, or quarry face. In such many and varied situations, it is not clear who should take responsibility for the continuing duties of monitoring, for the long-term stability of the 'asset', and ultimately for remediating any problems it may experience.

My suggestion, therefore, is that when constructing retaining structures to stabilise a landslide that involves two or more properties, that the elements of the scheme are clearly separated so that there is no doubt what owner owns what, or that the shared structure becomes part of a party wall agreement. Covenants may be required to prevent one owner destabilising a joint scheme at some point in the future.

My own particular *bête noire* is the coastal local authority that has a scheme constructed with the aid of a central government grant and then willfully neglects the need for maintenance so that the scheme becomes dilapidated. In the most egregious example I know of, there are major failures, all of which could have been 'nipped in the bud', but which have been allowed to deteriorate to such an extent that they will need extensive and expensive repair. No doubt another central government grant will be sought, and we taxpayers will foot the bill. Again.

GOOD DESIGN: A MIXTURE OF ROBUSTNESS AND ADEQUATE RESERVES OF STABILITY

Slope stabilisation designs need to be robust, in that underperformance of one element in the scheme does not automatically lead to collapse. This objective is achieved in part by not selecting a primary stabilisation method, something that can fail due to neglect or lack of maintenance, corrosion, accidental damage, *etc*. For me, this rules out drainage as a main method, but of course, drainage as a component of some other scheme is always useful.

DESIGN CODES AND SAFETY FACTORS

Traditionally, Factors of Safety recommended by design codes for slopes in the UK were around 1.3, but the design code we know by the shorthand EC7 contains an instruction to seek a design with a mobilisation factor of 1.25. I have several objections to this diktat, including that for small slopes, 1.25 is rarely enough, and for really large landslides, it may be unachievably high. It is also a fact that the actual Factor of Safety achieved is a function of how the shear strength parameters and ground water pressures are evaluated.

In the US, guidance is given by a variety of organisations, including the US Bureau of Reclamation and the US Army Corps of Engineers, with different factors of safety required for different types and stages of construction.

The lower values for the target value of F demand that the analyst is sure of the design values for shear strength and water pressures, in continuum analysis, the other parameters as well.

NOTES

1 Bromhead, E. N. (1997) The treatment of landslides. *Proceedings of the ICE, Geotechnical Engineering*, *125* (2), 85–96.
2 Bromhead, E. N. (2005) Geotechnical Structures for landslide risk reduction. Chapter 18, in T. Glade, M. Anderson, & M. Crozier (Editors) Part III: "Management Implementation – Site and Regional Methods" of "Landslide Hazard and Risk". John Wiley & Sons, 549–594.
3 I have a book from 1867 that cautions against entering wells or shafts because of the problem of carbon dioxide (referred to as *carbonic acid gas*). It is a shame that this advice was not followed at the Carsington Dam site, where four workers died in a not particularly deep shaft from carbon dioxide, in that particular case probably resulting from a chemical process in which limestone in slope drains was attacked by pyritic mudstones. The old book goes on to explain about the carbon dioxide given off by beer brewing, and the risk of leaning over the vat and being overcome by fumes!
4 This practice is shown in a wonderful photograph in Terzaghi and Peck's book. It shows precisely the manual excavation being undertaken in a cutting on the line just south of the Sevenoaks Tunnel portal. A large number of workmen are digging out the slip, loading the excavated Weald Clay into barrows, and off-loading it into flat, open-sided, wagons, no doubt to be reused in new railway embankments elsewhere. At the time of his visit, Terzaghi was in the UK consulting on the Chingford Reservoir embankment failure and the coastal landslide at Folkestone Warren. Whether or not he took the photograph himself is unknown. The process is also described in the children's book *The Railway Children* written by E Nesbit and published in 1905.

5 I can't help but be reminded of the adage that 'He who sups with the Devil should use a long spoon'. I was introduced to this maxim in a short story written by the children's author Malcolm Saville over 60 years ago and never forgot it – the long-reach excavator being my 'long spoon' and the unsupported excavation the very devil.
6 The built environment and its sensitivity to settlement is exemplified by the town of Ventnor on the south coast of the Isle of Wight in southern England. The town contains a great number of houses that go back to the early 19th century and a network of roads. They are already subject to differential movements due to landslide activity. A major deep drainage system installed in the complicated set of landslides would only exacerbate the negative experiences of residents. In contrast, the deep drainage system in the Folkestone Warren landslide in the early half of the 20th century has been successful in preventing the biggest types of movement. It is crossed by a railway, but railways accommodate small movements relatively easily simply by repacking or 'tamping' the track.
7 My late father enthusiastically added lime to his garden as it 'improved the soil'. In doing this he was following the example of generations of farmers and other gardeners. Crushed chalk is also used for this purpose, and chalk was often used in the past to fill stabilising counterforts for railway cutting slides. I don't think my father ever consciously added cement, but he swore that sometimes his garden soils were like concrete!
8 Beles, A. A. & Stanculescu, I. I. (1958) Thermal treatment as a method of improving the stability of soil masses. *Géotechnique*, *8* (4), 156–165.

Chapter 10

Epilogue

THOUGHTS ON LIFE, THE UNIVERSE, AND GEOTECHNICAL RESEARCH AND PRACTICE

It is a sad fact that even when you dedicate yourself to a particular interest and specialism, like slope engineering in the present case, it inevitably takes up only a small proportion of your working day unless you are in the fortunate position of being a distinguished professor in an even more distinguished institution, like the one who once confided in me that he found that 10 lectures in a year got in the way of what he really wanted to do. At the time, I was doing about 15 a week, marking student work, filling in some of my spare time with doing calculations for him, and still producing my own research and consultancy outputs, so I wasn't all that sympathetic. He'd pretty much dodged all the teaching on my MSc course when I was a student, so it had been his way of life for a long time. And then for those of you who are geotechnical specialists in practice, it is rare for slope engineering to be the majority of your workload, as the other stuff, like foundation engineering, retaining walls, pavement design, or in some countries, seismic liquefaction assessments, often takes the biggest slice of the day and has its own issues that make the work far from mundane.

Over time, you discover what kit you need to do the job most effectively, and you assemble your own toolkit. I have several such toolkits depending on whether I am in the UK or abroad, and whether or not I expect to be in the field or the office, or lecturing. By 'tools', I include software as well as physical items, but I also include attitudes and approaches to the task in hand. These latter things are important, especially when it comes to investigating failures, because firstly, many failures have common features that you soon learn to recognise, but occasionally, they are the result of something unusual or at least not encountered by you before, and an open mind, a command of geology and geotechnical science, and an inquisitive nature, are mental tools that you need to employ.

As for an epilogue, this chapter is a compendium of things left over unsaid from the preceding nine chapters, or perhaps just repeated yet again for emphasis.

DOI: 10.1201/9781003428169-10

THE OFFICE TOOLKIT

I have stretched a point by calling it a toolkit, but a well-equipped office is one of the best resources one can have. I suggest that this includes a personal library, even if the organisation you work for has its own, and of course there are the unpublished archives of the projects you have worked on. If you have a job in a large organisation, you will find your desk space is rather limited, and you may not have much more than a shelf or a small under-desk filing drawer, if that, as some organisations have hot-desking. In academia, one may have a shared office or, better still, once a senior position is reached, your own office. I started writing this book during the coronavirus lockdown, and even the company-employed engineer or geologist can benefit from a home office, which may be as little as an alcove, a separate room or, as in my case, a (reasonably well-equipped, insulated, but far too congested) shed in the garden.

My primary tool of choice from that toolkit is a personal computer, my current version of which has dual monitors with high resolution and excellent colour rendition, and, by the standards of the first dozen mainframes I used as well as early personal computers, colossal computing power and capacity. I once paid nearly £4000 for a monochrome laser printer, but the one I use now prints in colour and double sided to better than 4x the pixel density, and the inkjet printer I use is even better and can render photographs, although in both cases the supplies are far pricier than the printer itself. If I was less interested in the environment, I think that the economic answer is to buy a new printer each time and forget about replacing the toner or ink. All that stops me from doing so is being fussy about the waste, and the fact that I can't lift the things to dispose of them!

The inkjet printer is an all-in-one which can scan and copy, and which could even send a fax, if anyone I knew still had a fax machine. I have a stand-alone scanner too, and here's the big secret: eventually I will have to throw away everything in the computing line, because the manufacturer will not provide drivers (software that makes the thing work) when it stops being a current model. This problem is acute with printers and scanners. It was a problem when the interfaces stopped being supported, but now that these peripheral devices rely so much on the parent computer, they become unusable with such things as an update on its operating system as well as the physical connection issues.[1]

Obviously, your computer has an operating system and various small peripherals like a mouse and a keyboard, but over the past few years I have used Microsoft Teams and Zoom, and they are improved with a webcam and headset – the latter also enables you to use dictation software. I expected that the COVID-19 lockdown would accelerate the adoption of both home and remote working, and lo, it came unto pass!

Then, there is the software. Much software is downloadable for free, but that which isn't free tends to be expensive, with no middle ground.

For slope stability and seepage analyses, I do depend on my own software, but there are excellent commercial alternatives. I use Microsoft Office with the MathType add on equation editor for my writing, Dragon Naturally Speaking for dictation, and CorelDRAW! for my illustrations. I have colleagues who swear by Adobe Illustrator or even AutoCAD in place of CorelDRAW!

Some software that from time to time I find useful is the Microsoft Image Composite Editor (ICE), which takes a series of photos and turns them into a panorama. ICE is better than the in-camera panorama function found on some cameras because the output image is not limited in size. An added benefit is that ICE is free as well as good at the job.

As for a library, I am a creature of the print book age, but I am gradually migrating to electronic formats.[2] Compared to the average consulting office space I have enormous lengths of bookshelves, but even so had to discard a lot of textbooks when I closed my university office. Mostly they were not much of a loss. For many years I had responded to requests as to which was the best textbook to buy (as an undergraduate is likely to own just one, if that), and I would always respond by telling a student that browsing in the library was best, and then only buying the book that you return to again and again when information is needed. It rewards the author in a small way, and means that you have the book to hand when you need it, as it will not be out on loan to someone else.[3]

Journals online are a good thing but expensive if you are freelance. I have the *Géotechnique* collection up to 2001 on CD, and it certainly takes up a lot less space than the paper version of the journals (2.5Gb on a tiny USB device being even smaller than six CDs in cases, which in turn is a lot smaller than 52 moderately sized books). The same goes for your archive of past work, some of which is in written form and not necessarily in your memory.

At one stage in my life I did a lot of drawing on an A0 drawing board using tubular-nib ink pens and lettering stencils. Mostly, nowadays I do everything on my computer, although recently I had to return to the old technology and brought an A3 drawing board out of retirement. I also have an A3 light box that allows tracing without tracing paper or film! The drawing board is useful to lay out photographs, especially stereo pairs of air photos, for which I have a simple stereo viewer.

Small calculators[4] with trigonometric functions are useful, but seem to have given way to 'apps' on the desktop computer or even mobile phone.

Finally, the one document you need to keep accessible and up-to-date is your CV. You may need it for a different job or for promotion, and you will need it when pitching for work. Indeed, I needed mine to add to a proposal for this book. It may also be the case that you need different CVs for different purposes, as what impresses a consulting client may be very different to what is required by a university or whatever is required to get you a job.

FIELD TOOLKITS

Perhaps the greatest benefit in doing fieldwork in a small country like the UK is that you can travel by car, and that means that up to a point, you can take everything you might possibly need. The car, of course, is another tool! Even when I am not intending to do fieldwork, I carry a basic safety kit in my car, including a first aid kit for a small group of people. Most first aid kits do not include an eyebath or sterile eyewash, as I discovered once[5] when I got a grass seed in my eye: mine does now!

The space in the car boot allows me to carry a variety of items of high-visibility wear: waterproof coats and work trousers in both colours: yellow and orange. Network Rail likes orange, and so too do some oil companies. You have to be careful to carry and wear high-visibility clothing, which is essential if working around machines on site, but then sometimes you need to work inconspicuously, and green, buff or even camouflage outer clothing may be preferred.

As for tools, I have several digging tools. A full-size spade or a mattock are good for heavy digging but take up too much space to be carried always. Instead, I have a two-piece ex-army trenching tool that I have found better than a camping-shop equivalent. In addition, I have a variety of blunt knives, such as a stiff-bladed putty knife, for fine work such as digging out slip surfaces from an exposure or in clay cores. As I rarely do rock work, I usually don't carry a geological hammer.[6] I feel that this is a tool for a proper geologist!

I use a hand-held GPS sometimes (*i.e.* rarely), and a compass-clinometer a lot. I have both a lightweight plastic Silva 15TD-CL and a 'clone' of a rather more expensive gadget. The clone is heavier than the Silva compass, but far cheaper. I also have a hand lens and an ordinary magnifying glass. I rarely need binoculars.

Travelling by air limits what you can take with you, both in terms of the absolute weight limits and what you are allowed to take in cabin baggage. A colleague once mentioned that boots are the most difficult item to replace at short notice when abroad if checked-in luggage goes astray and so he likes to wear his, but this guarantees that you will be asked to take them off to be x-rayed as you go through the security checkpoint. Boots need good ankle support and non-slip soles, with steel toecaps needed if you are working around machinery or with heavy tools.

My cabin baggage always includes a very basic version of my first aid kit with pain relief, some antiseptic wipes, my eyebath (but not the fluids), plasters, and a change of clothes plus my laptop, cell phone, and camera.

A major problem with fieldwork is carrying everything – a traditional rucksack is fine for hauling possessions on a hike, but is on the wrong side of your body for getting small items out for use and putting them away. If the walking and carrying alternates with using small tools, for example a compass-clinometer to measure discontinuity directions in rock outcrops, then a rucksack with front pouches (such as those used in fly-fishing) are particularly useful, and can assist with keeping your balance if scrambling up steep slopes or walking over rough ground such as a boulder field. A telescopic hiking pole also assists in this: mine is dual purpose and can act as a camera monopod.

In the past I have owned expensive (and heavy) cameras, and spent more on them than on the film, which is a mistake that it took time to realise. I am now disinclined to carry heavy photographic equipment, as the results from tiny cameras or even smart phones are adequate for most purposes. The cost of each additional shot is negligible, and unless you are a photographic maestro, take lots of images. You can even take video clips, but don't forget to provide a commentary, which reduces the need to take notes. On the other hand, a field notebook is essential, especially if you draw sketches as well as writing text. I find that notebooks ruled with squares help me draw sketches.

However, the use of a smartphone to take photos and videos or to take dictation probably runs down the charge in its battery rather fast, and therefore it is useful to carry a recharging device or a spare, especially if you use the phone to keep in touch, not least for safety reasons.

A supply of transparent plastic bags, labels, and felt-tip pens is useful for protecting samples. It can be frustrating if the identification is lost, so I normally double-bag the samples (so bags of at least two sizes are useful) with labels and also a description marked on each bag.

Now that one is forbidden to enter a trial pit it is sometimes helpful to acquire a long stick on site to point things out, but a laser pointer does the job even better, although a powerful laser pointer cannot be carried in cabin baggage when flying.

I have deliberately dedicated only a short paragraph to food and drink, sunscreen, safety things like eye protection and helmets, sanitary products and all the other paraphernalia of working on site and in the field that are not specific to slope engineering, but of course you should work safely, taking care of yourself, co-workers, and members of the public by staying alert and acting responsibly.

THE LECTURER[7] TOOLKIT

Mostly your needs will be met with a laptop or notebook computer or a tablet. I don't get on so well with tablets, as I prefer a keyboard, and a tablet with a keyboard isn't so much different in weight or size to a notebook computer. I well remember travelling with boxes of 35mm slides and overhead projector transparencies, although that era is thankfully over now. Some presenters just travel with a USB pen drive, and that usually works if you know that you are presenting in a fully equipped auditorium. In looking after other presenters, I have discovered that they sometimes arrive with notebook computers that are incompatible with the connections for the projector, and I have a bagful of converters. If I can travel to the venue by car, I take my own projector, and sometimes even a spare computer!

The downside of using only a USB stick comes with embedded videos that may be in a format that isn't supported on an auditorium computer, which makes playing a presentation from your own computer a far better proposition.

A valuable accessory is the combined laser pointer and remote control for advancing slides. The laser spot isn't always powerful enough against the bright display of the projector in a large auditorium. As no kind of laser pointer is allowable in cabin baggage, I have a remote control without a laser pointer for use when I fly with only cabin baggage. It usually means that I lecture while pointing fruitlessly at the screen at numerous points in the presentation, wondering where the red spot has gone, and whether or not I have a spare battery with me!

The lecturer toolkit is still a must even if you only have a 10-minute slot in a conference programme, or are making a presentation to a client or to a team on site. The latter is unlikely to have the space or the facilities to set up a computer and a projector if it is a small site, but big sites with office complexes almost certainly will have.

CAUSATION AND LIABILITY

Those various toolkits are simply aids to finding out what went wrong, and what to do about it.

One way of looking at the problem of who (out of a series of actors) actually caused landslide movements on a site is to apply the 'straw and camel's back' principle, accepting that the last action was the 'trigger' and everything that went before was a 'preparatory' factor. This is particularly hard when the last thing done was a vain attempt to prevent the collapse, and that collapse was inevitable. An example of this might be someone being blamed for the slide who is placing small amounts of fill in an effort to level up a subsiding slope, not realising that instability was developing, or to infill tension cracks so as to prevent the ingress of rainwater and thereby had 'filled at the head of the slope'.

I find the alternative logical principle that *'if it hadn't been for X, then we wouldn't have had Y'*, where Y in this case is a landslide as sometimes helpful. Certainly, with X (whatever it is) we might have had ground movements, but they would have fallen short of creating the sort of landslide and associated deformations that we have. This is particularly acute when there is excavation at the toe followed by a series of small slippages, as these may ripple up the slope, with each collapse progressively undermining what is above until they meet something that was perfectly stable in its own right prior to encountering the slide that was enlarging 'headwards'. Something similar occurs with coastal or riverine erosion.

A third way of looking at the problem is not the last action on the slope, but the delay between that action and the collapse. So, if we have excavation at the toe and later filling at the head, then the 'last straw' argument puts the blame on the upslope side. However, excavations reduce pore water pressures in clay soils, which can take years or decades to come into equilibrium with water inputs and outputs on a slope. In contrast, the placing of fill not only adds load that is destabilising in itself, but it generates pore water pressures in the ground that are also destabilising in nature. Thus, the worst effects of a fill are immediate,[8] and the worst effects of the toe excavation may take a long time to manifest themselves. You can discount the effects of filling as the true cause if the failure is not more or less immediate, because the destabilising effects of the fill diminish with time. The fill may well have allowed the failure to occur earlier, but the cut was ultimately the cause.

As if that wasn't enough, it is worth considering *where* the toe and head of the landslide lie in relation to the postulated destabilizing actions. In one case I looked at, the head loading was nowhere near the first failure but was incorporated as the head of the slide sapped back, but the toe broke out where the lower slope retention structure was lowest and weakest.

A related problem is the case where a cut slope or embankment appears to be stable, but heavy rain or some release of water shows that it can, and did, fail. It is worth being reminded that cut slopes contain suctions that the surface water infiltrates and dissipates, and its previous stability was only apparent while the suctions were in existence. Some compacted fills also show this behaviour, especially when they have a substantial clay content. High or steep slopes may be held up by comparatively small, and therefore easy-to-lose, zones of suction at the toe. This came as a surprise to Charles Gregory in 1841, but should not be a surprise today.

WORKING AS AN EXPERT

Offering opinions on the above sorts of issue fall into the job description of a geotechnical expert.

Acting as an expert witness is one of those activities that is mainly routine office work and meetings, with many litigants settling before a trial, sometimes through mediation or deciding that a trial will cost more than just paying up, whether or not they think they have a case. Giving evidence is incredibly stressful, and if you are of a nervous disposition this is not a good career choice. The opposition barrister is not the slightest interested in seeing justice done, and instead is motivated (and engaged) only to win his or her client's case. To that end they will attempt to undermine you. They may not be able to do this technically, so then they will try to make you uncertain of yourself – possibly by embarrassing you, or making you angry. If you have a quick temper, this is also a line of business to avoid.

However, it is usually important, even if there is no litigation in prospect, to identify what caused the landslide, as that may well point out a simple and cheap option for remediation, for example to stop digging, or stop water entering the slope.

I have been involved in a number of cases over the years, and you must remember that you are as likely to be 'on the side of' a claimant (plaintiff) as you are of a respondent (defendant). The older, less preferred, terms are in brackets. I use the expression 'on the side of' as a shorthand, because you should keep yourself aloof and be independent in your views. Good advice is to write the same report regardless of who you work for, although my experience is mixed: sometimes the client and his legal team will have no truck with an expert who doesn't toe the line, while other clients are far more sensible and prepare themselves to eventually lose, although hopefully minimising their losses in some way. This does mean that you will be released if you don't toe the party line and you will certainly meet experts who only do expert work, and therefore lose their independence and become additional advocates. I really dislike dealing with those sorts of people.

You will find that sometimes the machinations of some lawyers sail very close to the wind in terms of what most normal people would think of as honesty. Each party is only allowed one expert, but if they appoint an independent firm to investigate a problem, and then appoint an expert for the legal side, they are trying to squeeze in two – and sometimes they get away with it. Moreover, if you find yourself appointed by one party with two or more in opposition, you will find that they usually gang up on you in meetings. The judge or arbitrator actually decides what are the facts of the case on the balance of the evidence, expert and otherwise, and it is usually rather unclear as to what the real facts are (in terms of engineering and science) amidst a welter of other, often obfuscating, detail. It is easy for the judge to be swayed when there are multiple experts on one side of the case, even if some of them are unofficial.

Barristers also have a range of theatrical tricks at their disposal to make a witness seem confused, unsure of their evidence, or unreliable, and if you do this sort of work and get on the witness stand, they will all be deployed against you. One of the most egregious acts I have ever seen was to suggest that an engineer was some sort of crook because he had changed his mind over a calculation. This had been done in a scrupulous way by lightly crossing through the original calculation, and recording in a marginal note what had been done. The cross-examination went on for some hours, and the engineer (who had a history of cardiac problems) was visibly distressed by the procedure. The lead solicitor took me aside and advised me not to show my utter contempt for the barrister in question when my turn came to give evidence. I didn't need to. The crook even showed me a drawing that had technical defects in its representation of part of the problem and, after I'd identified them, tried to claim that it was my drawing. It was in the same colours. When I called him out on it, it was "*No more questions*". I should have asked the arbitrator to rebuke him, but the planned deception was plain enough to be a factor in losing his side the case.

Another mechanism for destroying a claimant who has limited funds is to withhold critical sections of the documentation so that the other party runs up costs. Or to employ really expensive legal and expert teams so that the risk associated with losing a claim is too great to pursue it. Or to delay so that a litigant becomes insolvent. In another case, the claimant had actually caused the damage himself. The list of tactics goes on and on.

Most of us are familiar with the adage that *there are lies, damned lies and statistics*, a saying that was popularised but not invented by Mark Twain. Indeed, Wikipedia points out that it may be a mangled version of another, legal, adage that *there are liars, damned liars, and experts*. I have found that on occasion, the latter is very true. Just in the same way that statistics are often deployed to sell an untruth, experts aren't always right, and sometimes are unethically partisan. Beware.

ON THE SUBJECT OF MAKING MISTAKES

'To err is human, to forgive, divine'. People make mistakes all the time. The real art is not to design something where one mistake is catastrophic or disastrous. Partly, that means that you should design something that is robust, or in other parlance, 'fail safe'. In the stabilisation of slopes that means not just relying on one thing, especially one thing that may lose its efficacy with time. So, if drainage is a big part of the solution, it not only needs to be maintainable, it actually needs to be maintained. Then, if for safety reasons drainage adits may no longer be inspected throughout their length, then at least they need to be inspected at the outfall to be sure that they are still discharging. It may be sensible to put in new drains periodically to make up for efficiency losses in the old ones.

Personally, I rather like the idea of having something else, for example, a toe berm and drains, so that if the former is washed away or the latter get clogged, the engineer has time to find out that something is going wrong in time to take further steps to prevent things getting out of hand. Infrastructure asset management with periodic inspection helps.

But how can a householder with a retaining wall or a slope do this? I think it is incumbent morally on the original designer to make the need for periodic assessment of the stability of such things part of the expectation associated with ownership, just as the owner of a car gets it routinely serviced.

In the context of litigation, or even of life in general, it isn't so much the mistake that is the problem, it's the bull-headed obstinate refusal to heed a warning or press on with something unsafe or plain stupid that is so often the cause of the real calamity.

MISTAKES I HAVE MADE

As my advice is to come clean about your mistakes, then perhaps it is only fair that I describe mine, and how I discovered them. The subject of this book being what it is, I am of course only confessing to the mistakes in regard to slope engineering and not in my private life! Fortunately, no-one was killed or injured by mygeotechnical mistakes, and no-one lost money (except sometimes I did): these are mistakes of understanding.

Early on in my career I believed, on the basis largely of my undergraduate studies, that the slip circle was the ideal shape for a landslide slip surface, with the belief scarcely modified much by the 'friction circle' and log spiral shapes that appeared in textbooks. When I encountered a compound landslide, it then seemed obvious that the underlying strong bed had prevented the landslide from sliding round that ideal path, and so it had had to follow the compound shape. Some colleagues at the time thought it was one of Alexander Collin's log spirals, incompletely developed. With hindsight this is what one might call 'faith in low-grade soil mechanics'.

Later on, I was introduced to the idea that the bedding-related shearing was the result of the soil above expanding laterally more than the soil below with shear strains concentrated in a small zone. I would now call this 'faith in higher-grade soil mechanics'.

Of course, both the low- and higher-grade soil mechanics explanations recognize that there is something depositional in the soil succession without connecting it to where the slip surface developed. It took decades for me to realise what a slide prone horizon was.

Strong beds are easier to find in boreholes than weak beds, especially if the latter are very thin, few samples are taken and even those are not logged, and as a result, the former appear more easily in ground model diagrams.

I think that sometimes I have made mistakes due to believing what I have been told without questioning it. One of the bits of theory that, on reflection, I think has been mistaught is anything to do with geology-free soil mechanics, but then I've discussed this elsewhere in the book as well as in what follows.

A little bit of theory I do think relates to a misinterpretation is the moisture content cusp that is exhibited in slip surfaces. If a slip surface is exposed, particularly one associated with a particular slide prone horizon, the clay gouge certainly feels more plastic than the surrounding material. The experimental determination of moisture contents in the surrounding clay, both above and below the slip surface, shows a 'cusp' or distribution with highest water content in the slip surface gouge. A conventional explanation for this is that the clay dilates when it is sheared, pore water pressures in the slip surface reduce, and then water is drawn in from the surrounding clay. On careful thought, while the mechanism is plausible, and perhaps dilatancy on shearing may be a factor in slowing down the movement rate in cut slope failures, the classic cases of Henkel[9] were too fresh for the postulated mechanism to have occurred, and the mudslide cases of Hutchinson[10] were, in any case, bottoming out on a slide prone horizon, incidentally the same as adjacent compound landslides, and what was probably the case was that these bedding-related features had a subtly different mineralogy, and therefore a higher initial equilibrium moisture content than the surrounding clay, and the moisture content cusp was the result of geology and not soil mechanics! Taken further, where the measurements have been made, there is probably also a *smectite* cusp, and a lower initial shear modulus due to the higher water content, which is why the particular horizon was slide-prone in the first instance.

Landslide slip surfaces do not necessarily occur in the soil horizon that has the lowest residual strength; they occur where there is a low peak strength and mechanics that concentrate the strain there. Doing many residual strength tests on randomly selected parts of the succession is therefore as likely to underestimate the field strength of a landslide slip surface as to overestimate it.

PET HATES

I could probably fill a whole book with things I dislike for one reason or another, but for the present, I'll confine myself to matters that lie within slope engineering, and some of the ways that people handle the problem.

The first of these is the approach of correlating things and then claiming that the correlation implies causation. Correlation does not necessarily imply causation, but without correlation the cause-and-effect relationship *is* disproved. The folks who most love the technique take some simple factors, such as slope angle or soil type, and draw maps of each, and where the factors combine, they mark the area as red. Sometimes the factors are weighted before being combined. I'd always had misgivings about this approach, but they were reinforced by the mayor of a small town in southern Italy threatened by movements of a landslide hundreds of metres upslope and the possible release of a large landslide 'pond'. He knew precisely what the hazard was, and the risk it posed, including the 'where' as well as the 'what'. His comment was that he didn't want to spend money on getting a map with red areas on it; he wanted someone to take the problem seriously and do something about it. They still haven't.

I am often asked to review what I now call red area papers for a variety of journals. Sometimes they feature an elaborate statistical treatment, but come to the conclusion that only one factor is significant: in one case, merely the slope angle.

Another pet hate is to do with correlation between difficult and expensive-to-obtain parameters and something that can be measured simply and cheaply. So take, for example, a graph of the residual angle of shearing resistance *versus* plasticity index. Or maybe liquid limit. Or the SPT 'N' value. It doesn't matter. There will be a general trend, as low-plasticity soils generally have a low clay content, and therefore a high residual, but the bigger the clay content, the more likely the soil is to develop a slickensided shear surface, and the more likely it is to contain some of the slippier clay minerals. What I particularly dislike is the use of the line that someone drew on the graph as a definitive relationship, so that all that is required is to measure the plasticity, then derive the residual strength from the line. Usually, this completely ignores the scatter in the source data, and the fact that some of it is of better quality than the rest.

There are other correlations that must be true in general terms but which have a non-unique relationship, such as SPT 'N' *versus* undrained strength, or Young's Modulus, or some other parameter. In my view the correlations do not have this value, but they are useful in determining if your particular soil is run-of-the-mill, or if it is one of the outlier cases that could give you some nasty surprises.

The use of correlations this way is one facet of what I call 'geology-free' soil mechanics, and if more than 50 years in the business has taught me anything, it is that the geology has a huge influence on the behaviour of slopes. It does in tunnelling, and probably in foundations as well.

Another disconcerting thing is the failure to appreciate the influence of preliminary assumptions on the results of an analysis. There is an example of this in a method sometimes used in back analysis, which is to locate the head and the toe of a landslide on a cross section, then assume (a) that the soil is uniform, and (b) a particular level for the piezometric line, followed by the analysis of a range of slip circles that pass through those extreme points. Almost inevitably, there is a critical slip surface found. In many cases, the back analysis of the actual slip surface, found by difficult ground investigation, will yield a broadly similar back-analysed result for the angle of shearing resistance with *that particular piezometric line*, but what will never be indicated properly is where the actual slip surface is located, as that is controlled by the geology and geological history of the site.

Then, the assumed piezometric line gains the status of something that simply must be correct. If it had been assumed to be high, then the immediate result is to persuade the analyst (who, meanwhile, has not moved from the chair in front of his or her computer and has never visited the site) that drainage is the obvious solution to the problem. It might be, but it might not be. If the landslide moved under a low pore water pressure regime then the drainage may be ineffective. It certainly will be if the resistance is wholly or partly gained by a shear-key effect, and the drains did not go through the basal slip surface because in actuality it was deeper than indicated by the slip circles. Even worse, if the drainage is inundated before it is complete, and perhaps not properly connected to outfalls, it will act in reverse, and let water into the slope rather than removing it.

I also dislike the method of evaluating a balance between cohesion and 'friction' on the basis of slip circle analyses and matching the average or maximum depth of the indicated slip surface with the actual one.

Another pet hate is to have boreholes drilled with samples that are widely spaced. It is difficult enough to find slip surfaces in high-quality samples or cores, but leaving unsampled gaps in the succession turns a difficult job into an almost impossible one. I think that the procedure has its origins in investigations for piled foundations, where 'geology-free soil mechanics' is partly in operation. The assumption will be some gradual increase in strength with depth, where any strong horizons are an uncounted bonus and weak bands have little adverse influence, as they can be countered with more, longer or wider piles. This philosophy also deals with the scatter of test results by fitting a 'design line' through the middle. It does not work when dealing with a slope stability problem!

Not, of course, that I'm impressed greatly by any engineer who swears that their analyses are perfect, whether from the perspective that the analytical method chosen or the parameters used are 100% correct; they never can be. Or that the ideal Factor of Safety F is 1.3 or 1.25, and slavishly follows some practice that has been shown to be misleading.

I also have a huge number of pet hates to do with instrumentation. One of those dislikes relates to inadequate reading frequencies or durations, which mean that not only are you unsure as to whether equilibrium has been reached as the instrument settles down after installation, but you have no idea what the natural variability or extremes might be, and I resent the waste of money spent on them if they are unused. I also include in this category the failure to establish the baseline for inclinometers or the initial response curve for piezometers, both of which impair your understanding of what the current readings actually mean. I have sometimes been party to this error myself, in one case[11] considering the landslide's rate of movement to be so slow that a correspondingly long period between readings was called for, whereas the actuality was that the landslide was moving a lot faster than anyone (including me) realised, and the tube was sheared off before enough readings had been taken. Fortunately, there was also a small geodetic network and a GPS reference station on the landslide so that it was a lesser loss than it might have been.

The issues with laboratory testing are many. I particularly dislike testing without any obvious purpose, and I dislike the concentration of work on inappropriate tests. For example, in the Carsington Dam first construction a huge number of large-diameter hydraulic oedometer or 'Rowe cell' tests had been carried out, but only a half-dozen, small diameter triaxial tests. Now, an issue with those tests was that they had been misapplied to the different zones of fill in the dam, but far worse was that they had been done using a multistage procedure, where, after a point of failure has been established, the cell pressure is increased and shearing recommenced to establish further point or points on the failure envelope. This procedure ignores the strain-softening that occurs post peak, and rotates the failure envelope. While the loss of a degree or two in the slope of the envelope would make for a more conservative design, another factor comes into play, and that is where cell pressures have been chosen that give far more than field effective stresses, an artificial cohesion intercept makes the resulting parameters extremely unconservative.

Even when there is a comprehensive programme of strength testing, in a particular project, many of the tests will eventually prove to be of little relevance, and I think that I have equal dislike for the idea that strength testing on random samples is a good strategy, and for the criticism in hindsight that certain (expensive) parts of the testing programme demonstrated lack of judgement. Even though that is true, up to a point, judgement is always better in hindsight than it is under pressure at the time when the tests need to be specified.

And finally, I have lost count of the times when I've been told by someone with a landslide on their property that is weeping water rich in *e. Coli*, "*Oh no! It can't be our septic tank. We don't use much water, in fact we don't even flush the toilet every time* . . ." In the background, the plumbed-in washing machine whirs away, and enters yet another rinse cycle.

ONCE SEEN, NEVER FORGOTTEN

When you write about the things you *might* find, the observations you *might* make, or the analyses you *might* do, it sometimes comes over as though you *will* always find those things, you *will* make those observations, and you *will* certainly do those analyses. You won't. You certainly won't find all of them in every project, and whether or not you find some of them depends on how similar the landslide is to landslides you studied or investigated before. Some things you only see once, and never in your career will you see the same thing again. You will have to compile your own list, but here are some examples of things that I've only seen once or, very occasionally, a couple of times.

I'd often seen gypsum crystals when I've been logging in the upper part of the London Clay Formation. Many of the fossils in the un-weathered clay are preserved in the form of iron pyrites, and there is a lot of calcium carbonate, which is often found associated with the septarian nodules which occur in layers. Once the pyrite oxidises it releases sulphates, and inevitably the result is a change in colour from blue-grey to brown in the clay and the growth of gypsum crystals, which are calcium sulphate. Not all clay strata experience the same degree of colour change on oxidation as happens in the London Clay Formation, but it is a very obvious feature in this material. For many years I had seen quite large crystals of gypsum, but on one particular job it was obvious that the crystal growth had taken place along joints and fissures in the clay and also along what were obvious slip surfaces that had become stationary for any length of time. I've never seen it quite that clearly anywhere else.

However, once seen, the effect was never forgotten. It possibly accounts for the phenomenon of strength regain in slip surfaces that, for one reason or another, have become stationary for a length of time: the regrowth of crystals disrupts the alignment of the clay particles. Of course, if the reason for the landslide stopping movement is that the groundwater table lowers, then the slip surface becomes over consolidated, and that may account for some of the strength regain.

Again in the London Clay Formation, I have often seen those tiny and extraordinarily thin silt horizons or 'partings'. The core often breaks along them. It is difficult to see how the silt is distributed over large areas far from the mouths of rivers that bring in the sediments, but if you accept that they might be a small amount of fine volcanic ash, then the mechanism that imports it into the sedimentary record is a simple one. I've only once seen a thicker layer that is definitely identified as volcanic ash at the base of the London Clay Formation at Walton on the Naze in the coastal cliffs there. I'm sure that they are much more widespread than my experience. Seeing the Walton ash band was one of the important observations that gave rise to the content of my Glossop lecture.

Of course, shearing and landslide slip surface formation can occur anywhere in a clay deposit. It just happens that in the UK, it is extremely common for slip surfaces to develop along a susceptible horizon in sedimentary deposits, or what John Hutchinson and I termed a slide-prone horizon, because it isn't just weakness that causes the slip surface to form where it does, but other factors as well.

A phenomenon that I saw once in Java has stayed with me. A landslide had occurred during construction works at the foot of the slope to create a canal cooling water for a power station complex. The slope was actually an ancient lava flow, perhaps a hundred thousand years old, which had overrun a mangrove swamp and caused shearing at depth in the mud. The irregular settlement pattern as the mud consolidated gave the interface that sort of combination of humps and depressions that was reminiscent of an egg box. There had been a question as to whether the old mangrove swamp clays had been sheared prior to the landslide occurrence, and my very brief site visit did not reveal any signs of those shears. I stayed overnight, but the following morning before returning to the airport I decided to go for one last look, and there, beautifully exposed as the excavation works took place, was the slip surface, which I was able to sample, and then very quickly it was excavated away. I had never seen a slip surface generated by a lava flow into a mangrove swamp before and I've never seen one again, but they do exist. It was not a section that had slid in the later landslide, because that had all been excavated away, but was a relic of the original lava flow.

SEEN MANY TIMES, AND ALSO NEVER FORGOTTEN

Among the first few landslides that I worked on when I was in industry, one was a very pronounced graben-type landslide. I see these graben features again and again when working on landslides, and they fall into at least two categories: those that show the acute change in direction inherent in a compound landslide, and those that are really just tension or 'pull-apart' features at the head of a landslide which just get wider and wider as the landslide moves. A correlation was once published between the width of a graben and the depth to the slip surface – they are very approximately equal in many cases – but like most apparent correlations there are many outlier data points. It is also the case that landslides in different stages exhibit a changing relationship, and a compound landslide graben is inevitably rather different in its shape to a pull-apart one.

I wish that I'd realised earlier in my career that so many of the landslides that I would work on, even if thankfully it wasn't all of them, would be of compound shape with a bedding controlled slide-prone horizon forming the 'sole' (or the 'soul'?), and of very similar shape, because the residual angle of shearing resistance was always somewhere about 10–11°. The reason? See my Glossop lecture.[12]

FINAL THOUGHTS

What ultimately affects an engineer's ability to solve a landslide problem may, on occasion, just come down to a question of what can or cannot be done in a particular case, and which of a set of alternatives is most effective or cheapest. However, perhaps the majority of cases will never be dealt with by stabilisation because the problem is too big, the solutions too expensive or environmentally damaging, or sometimes because the landslide has happened and ends up in a state where engineered stabilisation is no longer necessary. Even a geotechnical career during which slope engineering forms only a small part of the workload will, sooner or later, provide examples throughout the spectrum of situations and responses.

Always be aware that sometimes the cure can be worse than the complaint, particularly if it is an attempt to construct an engineered remedial scheme that fails, and the failure causes mayhem. Once, some decades ago, after a discussion of that possibility, a friend and colleague articulated it[13] along the lines of:

> "*Imagine the situation. There has been unseasonal rainfall, and the landslides have gone on the move. Roads are torn apart, and services fail. Houses are collapsing, and people are rushing out carrying children and meagre possessions, screaming for help. The media are reporting it,*

with TV, radio, the Internet and the newspapers going mad. Questions are asked everywhere from Local Government all the way up to Parliament. Just remember, this might be the scenario if we do nothing. But equally, it could be the scenario if the remedial works go badly . . ."

It could be a topic for a novel. Someone may write it. It could even be me.

NOTES

1 At the time of starting writing, I had just transferred to a new computer, my old favourite having become prey to a failure in Windows 10 Update 1903, and finding that it wouldn't run, rolling back to the previous update repeatedly until locations on the solid-state drive wore out. The new computer has neither a parallel printer port nor an RS232 serial port, so even if I wanted it to, my plotter (stored away in my garage) wouldn't work. Nor would several printers, scanners and assorted other kit. Spilling tea on the keyboard also isn't a good idea.

2 It is even possible that you are reading this as an e-book. Don't worry, you are near the end!

3 I have lost some great books to students who borrowed them and then failed to return them.

4 They replaced the slide rule. I still have mine.

5 Twice, actually. Once on a ferry from the Isle of Wight, a week later in Ironbridge.

6 I even stopped these hammers being issued to civil engineering students doing geology field trips, as the blighters hammered anything they could find: stone buildings, stone walls, rock outcrops (but only occasionally) and sometimes even each other.

7 I mean doing lecturing occasionally, and away from your normal place of work. Lecturing in a university or college as a job is a different matter. I still, even in retirement, hanker after those early days when I could saunter into a lecture room armed only with a stick of chalk, or later, an OHP pen. Those were the days. And the approach was far more versatile than the ossified 'death by PowerPoint' that seems to have become the norm.

8 The authority for this statement being either Bishop and Bjerrum in 1960 or Hutchinson and Bhandari in 1971, and so it is a long-established principle.

9 Henkel wrote several papers on failures in cut slopes on railway lines in the London area during the 1950s. Why he failed to invent the concept of residual strength is a mystery.

10 Hutchinson described mudslide/mudflows on the north coast of Kent, in the London Clay Formation.

11 This case being that of the landslide at St Catherine's Point on the southern tip of the Isle of Wight. The GPS reference station was on the lighthouse, which moves about 100mm seawards along a very gently dipping slip surface at depth in a wet winter. (Hutchinson, J. N., Bromhead, E. N. & Chandler, M. P. (2002) Landslide movements affecting the lighthouse at Saint Catherine's Point. *Isle of*

Wight Conference on Instability: Planning & Management, Thomas Telford, May 2002, 291–298. The paper complements an earlier paper: Hutchinson, J. N., Bromhead, E. N. & Chandler, M. P. (1991) Investigations of the landslides at St Catherine's Point, Isle of Wight. *Slope Stability Engineering – Developments and Applications, Thomas Telford*, 169–179.)

12 Bromhead, E. N. (2013) Reflections on the residual strength of clay soils, with special reference to bedding-controlled landslides. *Quarterly Journal of Engineering Geology and Hydrogeology*, 46, 132–155.

13 Yes, Geoff, you will know it means you when you read this. I have paraphrased and elaborated on what you said, but it remains true, nonetheless.

Printed in the United States
by Baker & Taylor Publisher Services